Women in Agriculture & Rural Development

NIPA® GENX ELECTRONIC RESOURCES & SOLUTIONS P. LTD.
New Delhi-110 034

WOMEN IN AGRICULTURE & RURAL DEVELOPMENT

Editors :

Shakunthala Sridhara

B. Nagachaitanya

A.K. Chakravarthy

M.K. Nagamani

T.K. Prabhakara Shetty

NIPA® GENX ELECTRONIC RESOURCES & SOLUTIONS P. LTD.

New Delhi-110 034

NIPA® GENX ELECTRONIC RESOURCES & SOLUTIONS P. LTD.

101,103, Vikas Surya Plaza, CU Block
L.S.C. Market, Pitam Pura, New Delhi-110 034
Ph : +91 11 27341616, 27341717, 27341718
E-mail: newindiapublishingagency@gmail.com
web: www.nipabooks.com

For customer assistance, please contact
Phone: + 91-11-27 34 17 17
Fax: + 91-11- 27 34 16 16
E-Mail: feedbacks@nipabooks.com

ISBN: 978-93-58870-37-4

Composed and Designed by NIPA®.

For Inspiring and Encouraging Generations of Scientists to Excel and Provide Solutions to Favour's Problems

Preface

Women in India number nearly 500 million constituting 49 % of total poulation of which 70% live in villages. Farm women carry out 75-80% farm work with 50-66% of them contributing to agricultural labour. Raising nurseries, transplanting, manual weeding, harvesting, threshing and storage are exclusively women's activities. In case of livestock more than 90% of the work is done by women. In addition rural women manage household chores, bear and rare children. Some of the problems faced by them are scarcity of resources, drudgery, economic deprivation, gender bias in wages, denial to make decisions and use income generated by farm activities. They have limited opportunities to articulate their needs and concerns during planning process in the local bodies because of men's dominance.

A cursory review of problems of rural women reveal constraints, limitations, biases, mostly related to their gender. Some of the major areas affecting their life and hindering the realization of their potential for high economic growth and personal empowerment are:-

Traditional perceptions and attitudes which perpetuate gender biases and discriminate against rural women's access to community and household resources. At the same time degradation of natural resources in the process of economic growth is threatening women's access to subsistence livelihood resources in rural areas. Globalization could further marginalize rural women and result in greater inequity in accessing resources. Lack of or poor education adversely affects her ability to access the new technologies emerging in agriculture and information sector.

Mechanization and modernization of agriculture have led to increased agricultural productivity and decreased drudgery but have adversely affected the rural poor women by reducing demand for labour. Further, mechanization has occurred for activities usually carried out by men; women continue to toil in labour intensive chores. Introduction of weedicides and pesticides has deprived a large number of women labourers of their meagre source of income.

The provision of collateral free-group guaranteed micro credit has provided an important source of capital for rural women. These group based micro credit approaches have served the short term financial needs of rural households well, improving cash flows, benefiting women's livelihood, leading to social solidarity and changing social status. However there are negative effects by way of increased social conflicts in the community and the tendency of men to control women's access to economic assets as well as women being used as the front for control over loans or financial decisions.

Usually regional economic growth is accompanied by a steady reduction in the relative share of agriculture to Gross Domestic Product (GDP). The decreased agriculture affects rural women adversely by way of loss of livelihood, unsustainable food production, increased dependence on cash economy to access food, search for alternate, viable livelihood, etc. With agriculture being replaced by export driven production, land resources get consolidated forcing subsistence farmer to sell his land. In the process both rural men and women become agricultural labourers from being farmers.

Information and Communication Technology (ICT) has great potential to inform and educate rural women about agriculture, health and nutrition provided they are user friendly, have gender responsive content and applications. However, there are barriers to increased use of ICTs such as inadequate infrastructure, high cost, limited capacity and content relevancy.

Very few studies have examined work-related injuries among farm women. Compared to this there is at least some data on pesticide effects. Still there is a need for understanding the gender implications of pesticide use. Because in addition to the general implication of pesticides in leukaemia, lymphoma, aplastc anaemia, soft tissue sarcoma and cancers of various kinds, a variety of reproductive health problems of women are linked to pesticide exposure. Very low levels of pesticides have been found to induce DNA damage, mental retardation and delayed milestones in children born to mothers exposed to pesticides.

Women professionals in agriculture are far less compared to other branches of science and technology with women comprising only 6.6% of total agricultural scientists. Agricultural extension is also male dominated with only 15% of world's extension experts being women. Most extension programmes ignore the women component and overlook the different role of and responsibilities of rural women compared to men. Both women scientists and extension experts, akin to their rural sisters are burdened with the twin responsibilities of professional responsibilities and family management.

Keeping the above problems as issues to be debated, a National symposium on women in agriculture was organized at the University of Agricultural Sciences, Bangalore during April 10-12, 2007. This book contains full length papers presented and wherever authors expressed their inability to send papers, abstracts of their lectures is included.

An endeavour of this magnitude is possible only with the help guidance, and financial support from many quarters. The organizers and editors are grateful to Dr. M. N. Sheelavanthar, former Vice Chancellor of University of Agricultural Sciences, Bangalore, Dr. P.G. Chengappa, Vice Chancellor and Dr. T. K. Prabhakara Shetty, Director of Research, University of Agricultural Sciences, Bangalore for the patronage extended, technical support made available and facilities provided. We are also grateful to the financial support extended to us by various funding agencies namely Jawaharlal Nehru Center for Advanced Scientific Research (Bangalore), Council of Scientific & Industrial Research (New Delhi), Indian National Science Academy (New Delhi) and Department of Biotechnology, Ministry of Science & Technology, New Delhi. We are thankful to Ethological Society of India for sponsoring the symposium. Finally we are highly indebted to the delegates, invitees and participants of the symposium.

Bangalore
22-05-2008 *Editors*

List of Contributors

Afifa S. Kamili
S.K. University of Agricultural Sciences & Technology
Shalimar Campus, Srinagar-191 121
Jammu & Kashmir

Anand, T.N.
Communication Centre
University of Agricultural Sciences
GKVK, Bangalore-560 065

Archana Sharma
Department of Food Science and Nutrition
College of Home Science
CSK HPKV, Palampur-176 062

Aruna Rana
College of Home Science
CSK HPKV, Palampur-176 062
Himachal Pradesh

Ashwini, A.
Department of Agricultural Extension
University of Agricultural Sciences
GKVK, Bangalore-560 065

Balakrishna Gowda
Department of Forestry and Environmental Sciences
University of Agricultural Sciences
GKVK, Bangalore-560 065

Basavarajaiah, D.M.
Poultry Department,
University of Agricultural Sciences
Hebbal, Bangalore-560 024

Bharathamma, G.U.
Department of Agricultural Extension
University of Agricultural Sciences
GKVK, Bangalore-560 065

Biradar, A.P.
Regional Agricultural Research Station, University of Agricultural Sciences (Dharwad), P.B.No.18,
Bijapur-586 101
Karnataka

Chakravarthy, A.K.
Department of Agricultural Entomology, University of Agricultural Sciences
GKVK, Bangalore-560 024

Chandish R. Ballal
Project Directorate of Biological Control (ICAR), Post Bag No. 2491
H. A. Farm P. O., Bangalore-560 024

Chinnaswamy, K.P.
Department of Sericulture
University of Agricultural Sciences
GKVK, Bangalore-560 065

Dahiya, D.V.
Directorate of Extension Education
CCS Haryana Agricultural University
Hisar-125 004, Haryana

Deepali Chauhan
Department of Family Resource
Management, Home Science College
C.S.A.U.A. & T., Kanpur-208 002

Deep Punia
Directorate of Extension Education
CCS Haryana Agricultural University
Hisar-125 004, Haryana

Devaki Jain
"Tarangavana", D-5, 12th Cross
RMV Extension, Bangalore-560 080

Dhakshayani, C.
Department of Microbiology
MGR Janaki College of Arts and
Science for Women, Chennai-600 028

Dhananjaya, B.
Department of Agricultural Extension
University of Agricultural Sciences
Bangalore-560 065

Eswarappa, G.
Directorate of Extension
University of Agricultural Sciences
Hebbal, Bangalore-560 024

Fatima, S.
Department of Sericulture
University of Agricultural Sciences
GKVK, Bangalore-560 065

Geetha, K.
Directorate of Extension
University of Agricultural Sciences
Hebbal, Bangalore-560 024

Gladwin Joseph
Ashoka Trust for Research in
Ecology and the Environment
No: 659, 5th 'A' Main Road
Hebbal, Bangalore-560 024

Guggari, A.K.
Regional Agricultural Research
Station, University of Agricultural
Sciences (Dharwad), P.B.No.18
Bijapur-586 101, Karnataka

Jagadeeswara, K.
Directorate of Extension
University of Agricultural Sciences
Hebbal, Bangalore-560 024

Jagagadish, N.
Department of Sericulture
University of Agricultural Sciences
GKVK, Bangalore-560 065

Jamuna, K.V.
Directorate of Extension
University of Agricultural Sciences
Hebbal, Bangalore-560 024

Jatinder Kishtwaria
College of Home Science
CSK HPKV, Palampur-176 062
Himachal Pradesh

Kadar Khan, H.
Directorate of Extension
University of Agricultural Sciences
Hebbal, Bangalore-560 024

Kaiser Jamil
Genetics Department
Bhagwan Mahavir Medical Research
Centre, Hyderabad-500 004
Andhra Pradesh

Kalpana, B.
Directorate of Extension
Bakery & Value Addition Centre
University of Agricultural Sciences
Hebbal, Bangalore-560 024

Katteppa, Y.
Department of Agricultural Extension
University of Agricultural Sciences
GKVK, Bangalore-560 065

Krishnamurthy, B.
Directorate of Extension
University of Agricultural Sciences
Hebbal, Bangalore-560 024

Krishnamurthy, D.
Directorate of Extension
University of Agricultural Sciences
Hebbal, Bangalore-560 024

Kumarswamy, N.
Department of Sericulture
University of Agricultural Sciences
GKVK, Bangalore-560 065

Lalitha, P .N.
Department of Political Science
Government Arts College
Bangalore-560 001

Leela Sahijram
Division of Biotechnology
Indian Institute of Horticultural Research, Hesaraghatta Road
Bangalore-560 089

M. Gowda
Ashoka Trust for Research in Ecology and the Environment
No: 659, 5th 'A' Main Road
Hebbal, Bangalore-560 024

Malhotra, S.R.
Department of Food Science and Nutrition, College of Home Science
CSK HPKV, Palampur-176 062

Manjunatha, B.N.
Directorate of Extension
University of Agricultural Sciences
Hebbal, Bangalore-560 024

Mushtari Begum, J.
Retd. Professor
Division of Home Science
University of Agricultural Sciences
Bangalore-560 065

Nagaraj, G. N.
Directorate of Extension
University of Agricultural Sciences
Hebbal, Bangalore-560 024

Narayana Gowda, K.
Directorate of Extension
University of Agricultural Sciences
Hebbal, Bangalore-560 024

Narayana Reddy, M.A.
Professor of Horticulture (Retd)
University of Agricultural Sciences
Bangalore-560 065

Narayana Swamy, B.K.
Department of Agricultural Extension
University of Agricultural sciences
GKVK, Bangalore-560 065

Nataraju, M.S.
Department of Agricultural Extension
University of Agricultural Sciences
Bangalore-560 065

Nira Ramachandran
UNICEF, New Delhi

Pandurangaiah, K.
Co-ordinator,
Watershed Programme
University of Agricultural Sciences
Hebbal, Bangalore-560 024

Priscilla Jebakumari
Stella Maris College
Chennai-600 086

Punia, R.K.
Directorate of Extension Education
CCS Haryana Agricultural University
Hisar-125 004

Pushpa, B.
Department of Economics
Bangalore University
Jnana Barathi, Bangalore-560 056

Raghavendra Murgod
Department of Agricultural Extension
University of Agricultural Sciences
Bangalore-560 065

Raghunandhan, K.L
Ashoka Trust for Research in
Ecology and the Environment
No: 659, 5th 'A' Main Road
Hebbal, Bangalore-560 024

Raghuprasad, P.
Directorate of Extension
University of Agricultural Sciences
Hebbal, Bangalore-560 024

Ramakrishna Rao, L.
Directorate of Extension
University of Agricultural Sciences
Hebbal, Bangalore-560 024

Ramappa Patil
Agricultural Research Station
University of Agricultural Sciences
(Bangalore), Kathalagere
Davangere Dt., Karnataka

Ranganath, M.
Directorate of Extension
University of Agricultural Sciences
Hebbal, Bangalore-560 024

Rani Arvind
Directorate of Extension
Bakery & Value Addition Centre
University of Agricultural Sciences
Hebbal, Bangalore-560 024

Rashmi, K.
Department of Sericulture
University of Agricultural Sciences
GKVK, Bangalore-560 065

Rekha Dayal
Department of Family Resource
Management, Home Science College
C.S.A.U.A. & T., Kanpur-208 002

Sailaja, A.
Acharya N.G.Ranga Agricutural
University, Rajendranagar
Hyderabad-600 050

Sapna
SRF, Indian Agricultural Research
Institute, Pusa, New Delhi- 110 012

Satya Bhama
Department of Food Science and
Nutrition, College of Home Science
CSK HPKV, Palampur-176 062
Himachal Pradesh

Shakunthala Sridhara
Emeritus Scientist (Rodent Control)
University of Agricultural Sciences
GKVK, Bangalore-560 065

Shilparani, M.S.
Department of Agricultural
Extension College of Agriculture
University of Agricultural Sciences
GKVK, Bangalore-560 065

Shipra Sood
College of Home Science
CSK HPKV, Palampur- 176 062
Himachal Pradesh

Shivayya, V.
AINP on Rodent Control
University of Agricultural Sciences
GKVK, Bangalore-560 065

Siddappa Setty
Ashoka Trust for Research in
Ecology and the Environment
No: 659, 5th 'A' Main Road
Hebbal, Bangalore-560 024

Sultan Ahmed Ismail
Ecoscience Research Foundation
Chennai- 600 041 *and*
Department of Biotechnology
The New College, Chennai-600 014

Suman Lata
College of Home Science
CSK HPKV, Palampur-176 062
Himachal Pradesh

Swarna Vepa
M.S. Swaminathan Research
Foundation, Institutional Area
Taramani, Chennai-600 113

Uma, R.
Department Agricultural Extension
University of Agricultural Sciences
Bangalore-560 065

Usha Ravindra
Directorate of Extension
Bakery & Value Addition Centre
University of Agricultural Sciences
Hebbal, Bangalore-560 024

Vanita Chetan
Directorate of Extension
Department of Agricultural Extention
University of Agricultural Sciences
GKVK, Bangalore-560 068

Varadaraju, G. M.
Directorate of Extension
University of Agricultural Sciences
Hebbal, Bangalore-560 024

Varshitha, C. N.
Directorate of Extension
University of Agricultural Sciences
Hebbal, Bangalore-560 024

Venkataranga Naika, K.
Department of Agricultural Extension
University of Agricultural Sciences
Bangalore-560 065

Vinay Kumar, R.
Department of Agricultural Extension
University of Agricultural Sciences
Bangalore-560 065

Yadav, N.S.
Directorate of Extension Education
CCS Haryana Agricultural
University, Hisar-125004
Harayana

Contents

SECTION-2: INTERACTIVE SESSION-FARM WOMEN AND SCIENTISTS

SECTION-3 EMPOWERMENT OF WOMEN

SECTION-4: CAPACITY BUILDING

SECTION-7: DRUDGERY AND GENERAL PERSPECTIVES

SECTION - 1
KEY NOTE ADDRESS AND PLENARIES

1

Linking Science to Economics: The Women in Agriculture Connection

DEVKI JAIN

There streams of thought are presented in this paper:

One, that rural women are not only the warriors, creative actors in the *agricultural scene, or rural areas,* but also the urban poor, the migrants , the homeless, the source of labour supply into the expanding urban spaces, including trafficking.

I would then argue [two] that this spread of the location of "rural women" is the phenomena that we should now look at, the connectivity between the old divisions of rural and urban, as it indicates the pressures that the current development paradigms operating in our countries, is putting on the RURAL ... and therefore that we should look at the macroeconomic policies that are in place, globally and regionally and nationally.

Then go on to suggest [three] that symposiums like this one 'National Symposium on Women in Agriculture' need to consider moving itself from what has become a convention, to show that rural women are in destitution and suffering and paying heavy bodily prices of fetching water and fuel for coping to keep their families afloat as food gatherers-move from such heroic moving description of rural and underprivileged women- to make strong propositions on macroeconomic policies globally, regionally and nationally.

Much has been said about the, "missing women", namely the millions of women who are either unborn due to sex selective abortion or who do not reach adulthood due to early mortality, the killing fields of discrimination. Here the argument mooted is that, the voice of women, as a lobby is missing in the current debates on some of India's strong initiatives on economic policy.

There have been interesting and popularly argued differences of opinion on the special economic zones (SEZs). Earlier it was the confrontation of the Finance Ministry vs. the Commerce Ministry, and then entered the voice and protest by the farmers–a *Gram Sabha*, which refused to permit the use of their land for a SEZ, lately some political party confrontations, and the voices of the people's movements; including one of the latest entrants, the mentor of Infosys, Shri Narayana Murthy

But where are the voices of women in this debate? Women are the major employees in the SEZs. For example, the Madras Export Processing Zone (MEPZ) at Thambaram, on the outskirts of Chennai. Set up in 1984, which became a SEZ in 2002 along with seven other Special Export Processing Zones (Kandla, Kochi, Visakhapatnam, Falta, Noida, Surat and Indore), currently houses over 100 factories, all of them 100% export-oriented. The MEPZ zone employs around 22,000 workers. About 70% of them are women, most between 18 and 30 years of age. They work 10-12 hours a day, without overtime, and get daily wages ranging between Rs.30 and Rs.70. To meet production targets, they are compelled to work harder and longer until they burn out or quit. But when interviewed, just like their sisters in Sri Lanka, they are grateful for this opportunity (Suchitra, 2005).

Further since the SEZs require large tracts of land, there is the question not only of farmers losing their source of livelihood but often households losing their food security, as home stead land and kitchen gardens often provide the last resort.

A recent report from Vidharba, where there have been 800 suicides by farmers in the last few years reveals that many of the women, from the suicide households are moving into prostitution as a regular source of income.

The eminent Latin American economist Lourdes Beneria (2001) has traced what she calls shifting the risk in which she proves through field based data, that risk is being shifted from the capital to labour, from the state to the worker, as the organization of production moves from the formal to the put out or contract: and the SEZ firms are absolved from the labour and other laws that prevail in the country

SEZ is only one of the initiatives. The Foreign Direct Investment in retail trade is another. Given that in retail trade, especially in the tiny sector of pavement vendors, tiny enterprises, including participants in what are called the traditional *haats* in rural areas, are women; and that they are particularly engaged in the two or three consumer goods that account for the majority of the expenditure of Indian citizens, namely food and clothing and that more women are being almost "mobilized" if not pushed into self help groups to produce all kinds of small consumer goods. It is surprising that the women's movement, which agitated for reservation and against rape and dowry and *sati*, is not protecting these violations of livelihood! Almost all analyses of the increasing flow of women into prostitution, child trafficking, are being attributed to lack of livelihood, and here is the potential assault, and yet women are missing from the debates!

The third current debate where women are missing as a lobby is in agricultural policy. For example in agriculture, the analysis offered is that demand for food grains is declining. Therefore the prescription is that there is a need to shift to processing of agricultural products as well as exporting them, shifting dependence on crops, getting private sector into the rural areas etc. The *mantra* with which the approach to the 11th plan (Devaki Jain, 2006) launches itself, is growth of GDP –the fact, that the sectors that are leading that growth, are not creating jobs, (Bhattacharya & Sakthivel, Internet) is acknowledged by the same paper. It is well known that it is this decline in incomes of the poor or the bottom deciles (Jalan, 1996) that has depressed the demand for food grains as well has been responsible for the acute hunger that stalks India, especially its women and children. But the fact that women predominate amongst agricultural workers and at the lowest end; and are holding up the rural economy with their energetic self help groups etc is not being used for voice on hunger and food security and its relationship to land rights and agricultural policy.

Another area, where women are missing, is the debate on labour legislation. There is much debate on moving from strong unionized labour to flexi labour to invite competitive Foreign Direct Investment. Again, considering that it is women who predominate in this insecure, informal economy and the examples provided by organization like SEWA (Self Employed Women's Association, Ahmadabad) that trade unionism is crucial for them to negotiate security, we do not see the voice of the women's movement in this area.

So, India's "Missing Women" in terms of the Amartya Sen's interpretation of this concept will increase, unless India's women are not missing from the political economy debates that are determining the future of India's "Missing Women".

Therefore I would like to argue that symposiums like this one, National Symposium on Women in Agriculture need to consider moving themselves from either the usual celebration that women are embedded in agriculture, that they are the food providers, the *kamadhenus*, need to be taught better processes, given opportunity for training etc, to understanding and articulating what is happening to the women in agriculture especially the women labourers, the poor households, due to the macro policies associated with a growth driven economy, where the growth pole is not well being , but export earnings, not employment or food security but export earnings. Forums like this one need to connect science to economics- to make strong propositions on macroeconomic policies globally, regionally and nationally.

Much has been written on why this is so: women worker's wages have been consistently and significantly lower than that of men and in some countries it is just above half those paid to male workers; women submit meekly to excessively long hours of work, hazardous and unhealthy working conditions; and they are not easily unionized. There is also the supply side argument, namely that women yearn to have a wage income and a regular wage income at that, rather than to be in the condition that they often are, namely hungry, ill and assaulted, with unpaid work. Getting out to work early in the morning and coming back however late, but with the joy of having earned monetary income certainly adds to self worth. Hence when women workers in SEZ enterprises are interviewed, they do speak about its positive value in their lives.

This trend towards feminisation of employment in Asian countries resulted from employers' needs for cheaper and more "flexible" sources of labour, and was also strongly associated with the moves towards casualization of labour, shift to part-time work or piece-rate contracts, and insistence on greater freedom for hiring and firing over the economic cycle. All these aspects of what is now described as "labour market flexibility" became necessary once external competitiveness became the significant goal of domestic policy makers and defined the contours within which domestic and foreign employers in these economies operated.

There are similar experiences from Latin America and East Asia- the early birds in getting investment through the SEZ or free trade zones as they were called, about the harsh if not enslaved working conditions, the "maquiladoras" of Latin America are well known . Further there is increasing information about the insecurity of these jobs. The uncertainties associated with the rejection of consignments leading to losses and losing of jobs, which are insecure in nature. There is an inability to reorient to export production and this, results in job loss and work under tremendously harsh conditions.

References

Bhattacharya, B. B. & S. Sakthivel. Economic Reforms and jobless growth in India in the 1990s. http://iegindia.org/worksakthi245.pdf.

Bimal Jalan. *India's Economic Policy – Preparing For The 21st Century*. 1996.

Devaki Jain. 2007. 'Reconceptualising inclusion: learning from Gendering public policy. To be published by *Yojana*, March 2007, *prepared for an approach to the 11th five year plan, Planing Commission*. Government of India, November, 2006.

Lourdes Beneria. 2001. Shifting The Risk: New Employment Patterns, Informalization, and Women's Work. *International Journal of Politics, Culture, and Society* 15 # 1/ September, 2001.

Suchitra, M. 2005. SEZs : Economic or exploitation zones? http://www.infochangeindia.org/agenda7_03.jsp (Suchitra won the Appan Menon Memorial Award 2005 for a study titled, *'The Socio-economic Impacts of the Proliferation of SEZs in India*)

2

Low Dose Effects of Pesticides in Peripheral Blood Samples of Donors and Farmers

KAISER JAMIL

ABSTRACT

In this investigation the low dose effects of pesticides were analyzed both *in vitro* and *in vivo* in peripheral blood samples from healthy donors and also in farmers occupationally exposed to pesticides. *In vitro* studies were conducted to assess the toxicity caused by commonly used pesticides like monocrotophos, profenofos, chlorpyriphos, endosulfan, acephate on lymphocytes from peripheral blood samples. In almost all experiments we could determine chromosomal aberrations and DNA damage at 1/10th of LC_{50} dosage of the pesticides. For analyzing the *in vivo* effects, epidemiological studies were conducted in pesticide exposed agricultural workers. The 200 exposed volunteers suffered from fever, nausea, headache and other abnormal symptoms and visited the hospital for general health check-up. These cases were taken up for more detailed studies. Five of them (2.5%) showed decrease in RBC, haemoglobin, and increase in WBC with a large number of immature cells. These volunteers were further diagnosed as Philadelphia negative Chronic Myeloid Leukemia (CML) cases based on clinical and pathological examinations. In similar environs, about 3% of the exposed children showed signs of mental retardation and delayed milestones. It is therefore, concluded that clinical evaluations supported by occupational epidemiology could determine CML in exposed individuals (adults). At low doses we have traced the effect of pesticides at molecular level.

INTRODUCTION

Organophosphorus pesticides which constitute the major class of pesticides along with the carbamate and organochlorine pesticides are widely used by farmers because of their high efficiency towards the target organisms. Extensive use of these pesticides in crop protection and for household purposes has resulted in their wide spread distribution in the environment causing part of the world wide environmental contamination. Besides their great contribution to controlling vectors of animal and human diseases, their use also creates many problems because of their toxicity to non-target organisms, persistence and combined effects with other pesticides and environmental factors. Despite the benefits, pesticides are known to have various detrimental effects on humans and their environment. Studies have demonstrated strong correlation between chronic exposure to pesticides and increased risk of developing various neurological, developmental, and genetic diseases (Table 1).

Table 1: Observations of children showing mental retardation

	A	B	C
1. Observation Length of exposure of mother during pregnancy (months)	3-4	6-7	4-5
2. Health Effects Type of exposure	Indirect	Indirect	Indirect
3. Adverse Effects observed	Normal	Normal	Normal
Karyotyping	No	No	No
Consanguineous birth	No	No	No
Genetic predisposition	Yes	Yes	Yes
Mental retardation	Yes	Yes	Yes
Delayed milestones	Negative	Positive	Negative
Heart impairment	Negative	Positive	Negative
Respiratory complications	No	No	No
Other abnormalities	–	–	–

During the sixties, it became evident that persistent use of pesticides was having an adverse impact on ecological communities. These studies faced numerous methodological problems common to environmental epidemiology, and even today, our understanding of the relationship between pesticides and human health is limited.

Occupational exposure to agrichemicals may be associated with adverse effects including birth defects, fetal death, reproductive failure, growth retardation and cancers.

Implications of pesticide toxicity as a cause of hematologic tumours in children including non-Hodgkin's lymphoma and leukaemia are now well documented, although the general public is still unaware of this. Some children have an overall increased risk of acute leukaemia if exposed to pesticides in-uteri or during childhood, especially exposure to insecticides and herbicides used on lawns, fruit trees, gardens, and for indoor control of insects.

Once pesticides enter the body system, the chemicals may produce various adverse health effects such as damage to the nervous, respiratory, immune, endocrine, and reproductive systems which may result in birth defects, cancer, and neurodegenerative diseases. Pesticides may affect the central nervous systems by inhibiting enzyme activities and suppressing neurotransmission to organs and motor nerves. Some symptoms of acute pesticide exposure include seizures, nausea, visual disturbances, increased heartbeat and blood pressure, depression of respiration and circulation, and death. Chronic exposure to pesticides also produces fatal effects including reduced motor functions, disorders of the immune system, neurobehavioral dysfunction, reproductive anomalies, and cancer. The neurological effects of pesticide exposure are usually cumulative. Researches have associated pesticides with neurological diseases such as Parkinson's disease, amyotrophic lateral sclerosis, and dementia.

In India, farmers make up about 70% of the population and consequently are the major targets of the hazardous effects of these toxic chemicals. Moreover, in developing countries the users are rather careless in handling pesticides. All pesticides in a given chemical group generally affect the human body in the same way; however, severity of the effects varies depending on the formulation, concentration, toxicity and route of exposure of the pesticide. It is important, therefore, to know the type of pesticide associated with poisoning.

The main objectives of the present study were to examine the health outcomes of a group of agricultural workers occupationally exposed to the pesticides in the State of Andhra Pradesh, India. Most of these exposed groups were apparently healthy; when they came to the hospital with small ailments; epidemiological data of such groups was recorded.

METHODOLOGY

Blood samples from 200 healthy adults, non-smoking, non-alcoholic and with little or no exposure to pesticides were collected as controls. An equal number of blood samples were collected from male and female individuals (adults) exposed directly or indirectly to farm or indoor pesticide sprays and who came to the hospital with complaints and disease-like symptoms (Table 2).

Similarly, blood samples from 100 children living in clean environs along with an equal number of samples from children living in pesticide-prevalent environments were collected in sterile vacutainers containing anti-coagulant for epidemiological studies.

Collection of blood samples and lymphocyte culture: Short term lymphocyte cultures using the whole blood samples were set up following the methods described by Yadav & Kaushik (1996) with slight modification. The simplest method of lymphocyte culturing is by incubation of a small amount of whole blood without previous separation of erythrocytes and granulocytes.

Trypan blue dye exclusion assay for Cytotoxicity: The *in-vitro* cytotoxic effects of various pesticides such as monocrotophos, chlorpyriphos, carbofuran, profenophos and dimethoate were estimated by the Trypan blue dye exclusion technique (Cook & Mitchel, 1989). Cells were counted using a Neubauer's chamber (Haemocytometer) and the number of viable (opaque) and non-viable (stained blue) cells were scored by direct observation under the bright field microscope and the results were subjected to probit analysis. All the experiments performed were in triplicates. For this assay, various aliquots of pesticide solutions were taken from the stock solution and added to the blood samples and incubated for 2hrs. These samples (treated) were stained along with untreated samples with trypan blue and the viability and mortality of the lymphocytes was recorded, as described earlier.

Probit Analysis: The cytotoxicity of the selected pesticides (e.g. monocrotophos, profenofos etc.,) was determined at various concentrations for lymphocyte mortality. The log concentrations and percent mortality data was subjected to probit analysis using 'the probit model program' (Reddy *et al.,* 1992) and lethal dose (LD_{50}) of the test compounds were calculated.

Initiation of cultures for determining chromosomal aberrations: Venous blood drawn from healthy non-smoking donors was used for the experiments. To culture the lymphocytes 1.5 ml of freshly collected whole blood was added with 2 units of PHA to each 5ml media vial prepared with autoclaved double distilled water having RPMI 1640 (5g/100ml), Sodium bicarbonate (1g/100ml), Fetal calf serum(10ml/100ml), penicillin (100IU/ml) and streptomycin (100IU/ml), maintaining a pH of 7.2 -7.5. The vials were incubated at 37^{0}C for 72 hrs. At the end of 48th hour of incubation, various aliquots of the pesticide solution were added to the tubes and incubated for another 24hrs. The tubes were shaken every morning until processed.

Determination of Chromosomal aberrations: Pesticide exposed cultures and non exposed controls were screened for chromosomal aberrations. All experiments for chromosomal aberrations were carried out in triplicates, and for each set 100 metaphases were screened. Metaphases suitable for cytogenetic analyses were screened for the frequency of chromosomal aberrations.

Analysis of chromosome aberration frequency: The protocol for chromosomal aberration analysis was adopted from Moorhead *et al.*, (1960). Chromosome preparations were screened after adding colchicine to arrest the cells in metaphase stage (after 72 hrs of initiation of cultures). These were fixed in methanol and acetic acid (3:1) and slides were flame dried and stained with 4% Geimsa stain before viewing under the microscope and the image was recorded in the Medi-image software program. The experiments were carried out using various aliquots from the stock solution of the pesticide. The concentrations used for the experiment were 1/10th and 1/6th of the LC_{50} dose in varying concentrations i.e. 12µl & 20µl of 1% pesticide solution for each sample. The lymphocytes were incubated with the pesticide solution for 2 hours. Untreated blood sample was taken as control.

DNA damage studies by Comet assay (SCGE): The methodology for the SCGE assay was based on Singh *et al.*, (1988) and was carried out with slight modification. Slides pre-coated with normal melting agarose (1%) were layered with low melting agarose (0.5%), along with 20µl of sample mixed in the agarose followed by another layer of agarose

(1%). Slides were immersed in cold lysine solution overnight at 4^0C. Prior to electrophoresis, the slides were equilibrated in alkaline electrophoresis solution (1mM Na_2 EDTA and 300 mM NaOH, pH>13) for 20 min. and electrophoresis was carried out in the same buffer for 25 min. at 25v and 30 mA. DNA fragments in each cell migrate at a rate inversely proportional to the size of the fragments. Slides were then washed gently 2-3 times, 5 min. each with 0.4 M Tris at a pH of 7.5 (neutral buffer). After the final wash, the neutral buffer was drained and washed with distilled water. Each slide was stained with 60 ml of silver nitrate (mg/ml), covered with a micro glass cover slip and sealed. These slides were washed three times with distilled water. The silver stained slides were viewed under a microscope, with a CCD camera attachment which was connected to a computer with Medi-Image software containing frame grabber, viewer saved in a library and finally a printable version was obtained.

Complete Blood Picture: ADVIA-60-Cell Counter (auto analyzer) was used to assess the percentage of haemoglobin, RBC and WBC count in all the 600 samples collected from adults and children. Differential leukocyte counting was done using Leishman's staining protocol. Thin smears of blood were spread on clean, non-greasy, dry slides and subsequently stained using Leishman's stain for 2 min. and washed with buffered saline for 20 min. The smears were washed, air dried, cover-slipped and observed under light microscope at 10 and 40x magnification. Blood samples of all individuals (exposed and unexposed to pesticides) were tested for ESR using Wintrobe's method. The rates of sedimentation of blood were recorded using the readings on the Wintrobe's tube. Bone marrow tests for diagnosis of chronic myeloid leukaemia (CML) were done on subjects showing abnormal results during blood testing. Bone marrow aspirates were taken to confirm CML morphology in the subjects.

Statistical Analysis of results was performed using analysis of variance (ANOVA) and the standard error was calculated for chromosomal aberration frequencies between treated and untreated samples.

Cytotoxicity testing: The viability of cells after treatment with pesticides for 2 hrs for estimation of the LC_{50} values was determined using the trypan blue dye exclusion method. Viable cells appear transparent, and dead cells are stained blue because of increased cell permeability. As the concentration of pesticides increased, the viability of cells decreased correspondingly. The percentage viability was calculated using the probit analyses software (Kaloyanova, 1985; Reddy *et al.*, 1992; Zurlo *et al.*, 1996). The percentage mortality caused by the pesticides individually and in equimolar combinations was estimated.

RESULTS

Pesticide usage: Pesticides used by subjects for household pests were Acephate, Chlorpyriphos, Chlordane, Dimethoate, Allethrin, Pipernoyl Butoxide and Transfluthrin. Those used on farm lands to control pests were Endosulfan, Chlorpyrifos, Profenofos, Monocrotophos, Carbofuran, Cypermethrin and Cyfluthrin.

Epidemiological studies conducted on pesticide exposed agricultural workers revealed that all the 200 exposed volunteers were suffering from fever, nausea, headache and other abnormal symptoms and visited the hospital for general health check-up. These cases were taken up for more detailed studies. Five (2.5%) showed decrease in RBC, haemoglobin, and increase in WBC with a large number of immature cells. These volunteers were further diagnosed as Philadelphia negative Chronic Myeloid Leukaemia (CML) cases based on clinical and pathological examinations.

Cytotoxicity of pesticides on lymphocytes: The individual treatments, of pesticides revealed the order of toxicity of the pesticides as: Profenofos > Endosulfan > Carbofuran > Monocrotophos > Chlorpyrifos > Acephate.

Chromosomal aberrations induced by pesticides on lymphocyte cultures: Genotoxicity assessment of pesticides at the tested concentrations of $1/10^{th}$ of the LC_{50} doses revealed that these pesticides could cause significant chromosomal aberrations in peripheral blood lymphocytes of humans in vitro. Metaphase plates with aberrations are presented in fig.1. It was found that, satellite associations increased with the increase in concentration of pesticides. All the data were statistically significant at a $p < 0.05$ using ANOVA.

DNA damage is idicated by the appearance of characteristic comets after subjecting the cells to alkaline electrophoresis. The morphology of normal cells and damaged cells is shown in fig 2. It is evident from the results that the DNA damage (comet tail length-µm) increased with the increase in the concentration of pesticides. Profenofos caused the highest DNA damage while acephate caused the least DNA damage. At concentrations above 3.5μM necrotic cells could be observed indicating enhanced DNA damage. All the data were statistically significant at $p < 0.01$ using ANOVA.

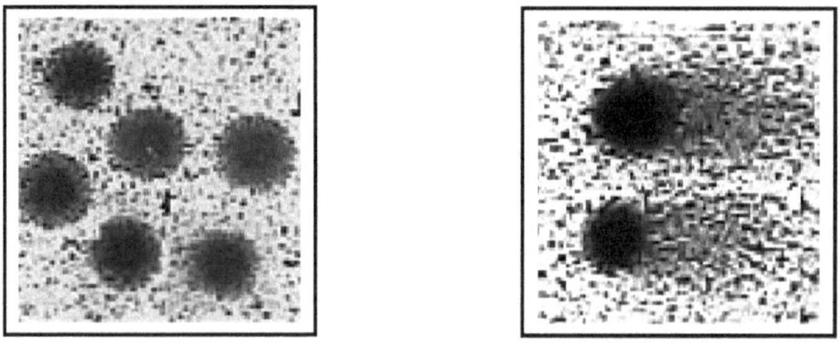

a. Control cells b. Cornets showing damage

Fig. 1. Chromosomal aberrations caused by pesticides.

The peripheral blood smear of the five volunteers who exhibited CML showed leukocytosis, with prominence of maturing myeloid precursors (Tables 3 and 4). Significant percentages of blasts and basophiles were seen. The bone marrow biopsy showed prominence of myeloid. There was significant damage in the individuals having hematological disorder. There was an increase of necrotic cells in these individuals. Of the 200 exposed individuals approximately five showed morphologic features of CML, other than genotoxic effects and therefore, we carried out detailed investigations on these particular volunteers.

The results are presented in Tables 2-4. No such effects were observed in controls and in exposed children. The incidence of CML-like signs in both direct and indirect exposure to pesticides in adults was about 2.5%.

Table 2: Epidemiological data of subjects showing symptoms of CML

Observation	Individuals showing CML				
	A	B	C	D	E
Age (yrs)/ Sex	34/ M	32/ M	35/ M	30/ M	28/ M
Kind of Exposure	Direct	Indirect	Direct	Direct	Indirect
Length of Exposure (days)	2	1	5	2-3	2
Genetic abnormalities	Nill	Nill	Nill	Nill	Nill
Neurological complications	Nill	Nill	Nill	Nill	Nill

Chromosomal aberrations

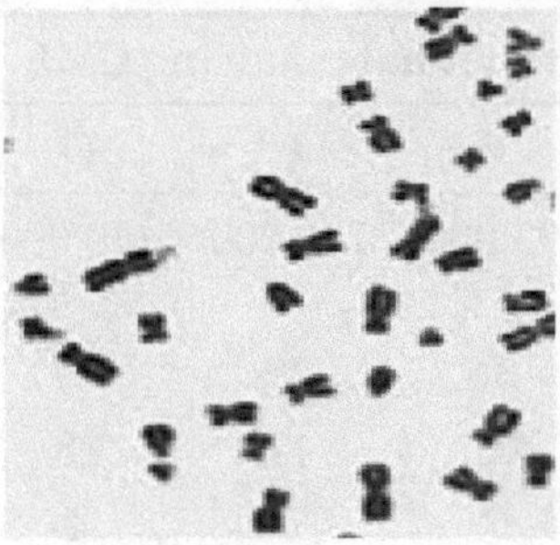

a. Control metaphase plate

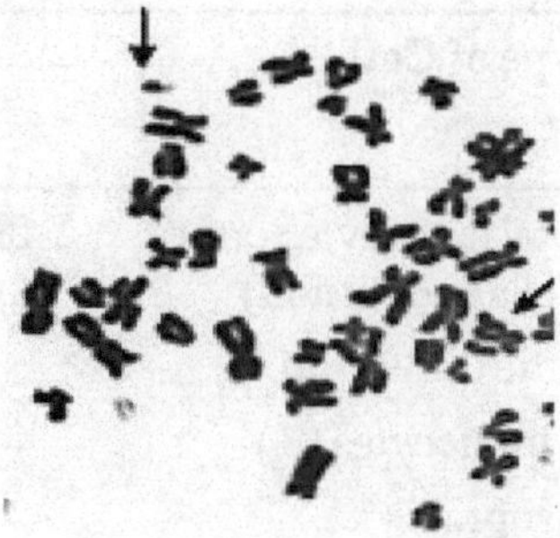

b. Metaphase plate showing chromatid break and fragement

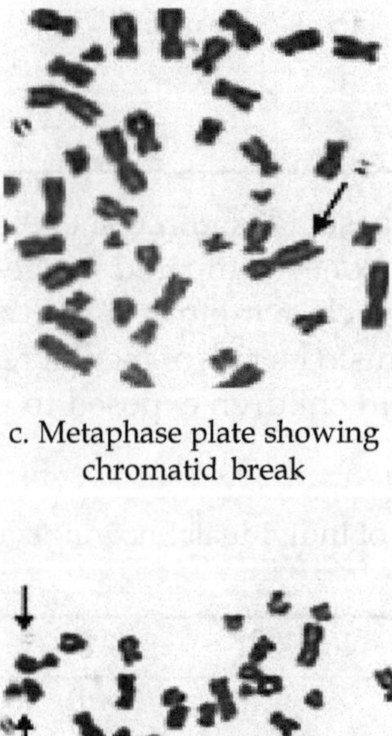

c. Metaphase plate showing chromatid break

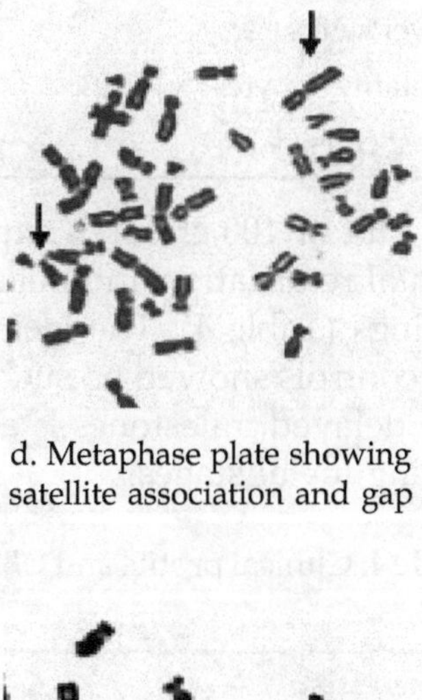

d. Metaphase plate showing satellite association and gap

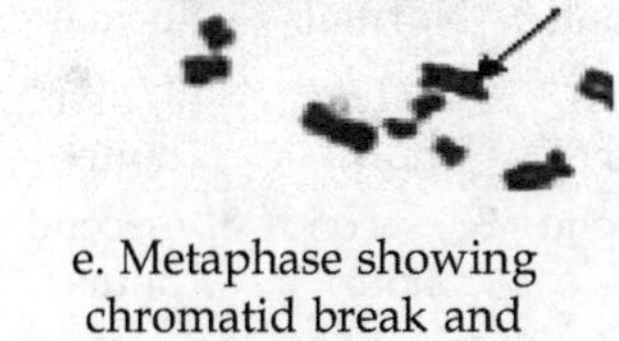

e. Metaphase showing chromatid break and association

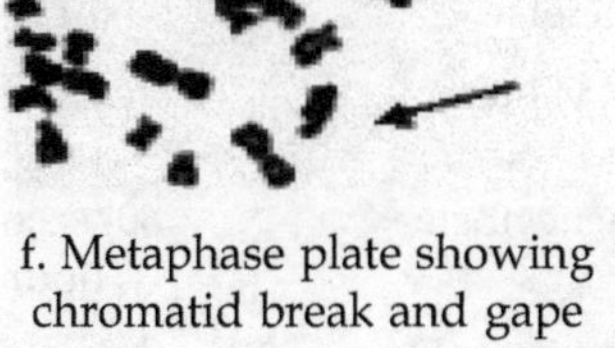

f. Metaphase plate showing chromatid break and gape

Fig.2. DNA damage (Comet assay) caused by pesticides

Table 3: Differential leukocyte count of WBCs of subjects showing symptoms of CML

Type of Cell	Individuals showing CML				
	A	B	C	D	E
Total leucocyte count (Cells/ cubic mm)	183, 800	93,000	230,000	130,000	92,600
Neutrophils (%)	18	36	39	29	52
Lymphocytes (%)	15	21	14	28	23
Basophils (%)	6	7	4	4	7
Eosinophils (%)	3	2	4	6	1
Monocytes (%)	4	6	3	3	2
Promyelocytes (%)	8	7	6	4	5
Myelocytes (%)	19	12	19	15	14
Metamyelocytes (%)	16	3	7	4	4
Blast cells (%)	11	6	4	7	2

Out of 100 children exposed to pesticides, three showed marked mental retardation and delayed milestones compared to age-matched siblings (Table 4). The incidence of such symptoms was 3%. All the 100 controls showed no such symptoms. The risk of mental retardation and delayed milestones is about 3% in children exposed to pesticides during foetal stages.

Table 4: Clinical profile and CBP analysis of Individuals Showing symptoms of CML

Observation	A	B	C	D	E
Haemoglobin (g/dl)	10	10	10	10	10
RBC count (million cells/cubic mm)	3	3	3	3	3
Platelets	Normal	Normal	Normal	Normal	Normal
ESR values	40/first hour	40/first hour	40/first hour	40/first hour	40/first hour
Leucocytosis	80/second hour	80/second hour	80/second hour	80/second hour	80/second hour
Splenomegaly	Negative	Negative	Negative	Negative	Negative
Hepatomegaly	Positive	Positive	Positive	Positive	Positive
Presence of Myelocytes	Positive	Positive	Positive	Positive	Positive

Out of the 200 cases which we examined (pesticides exposed), five individuals showed an increase in the total WBC counts (92,600-230,000 cells/cubic mm) compared to controls (4000–10,000 cells/cubic mm). There was also the presence of cells like promyelocytes, myelocytes, metamyelocytes and blast cells along with a decrease in the lymphocyte population (Table 3). An increase in the total WBC count accompanied a decrease in RBCs, elevated ESR values and alterations in platelet counts. Splenomegaly was also reported to have occurred due to hyperplasia and hepatomegaly in some subjects exposed to pesticides. However, there does not appear to be an inherited disposition.

DISCUSSION

Currently, about 153 pesticides have been registered and 35 are widely used in India. Of the registered pesticides, 33 have been banned, based on their persistence and toxicity towards non-target organisms including humans, both in India and in developed countries. Efforts are underway to avoid the use of such banned pesticides, and replace them with eco friendly chemicals. Malignancies and hematological disorders linked to pesticides have been reported in leukemia, soft tissue sarcoma and non-Hodgkin's lymphoma and also been documented in animal models. Risks from exposure to pesticides are especially high for persons with asthma, diabetes, and cardiovascular disease etc. These are issues of concern as they relate to environmental burden of pesticides and the exposure of humans to these. Pesticides are effective in killing pests through their neurotoxic effects. Infants appear to be particularly susceptible to the effects of these pesticides because they have incompletely developed acetyl cholinesterase systems and their immature livers cannot detoxify these compounds.Human peripheral blood lymphocytes were used for their predictive handling methods and uniform evaluation thus permitting sensitive detection of geno and cytotoxicity of environmental contaminants (Olive *et al,* 1993; Dearfield *et al* 1993; Jamil *et al.,* 2004). This provides a reliable basis to predict whether a chemical is likely to be harmful and if so at what levels of exposure or dosage. These results in human cell model give some insight into the different mechanism involved in cellular toxicity at very low dosages.

In India, 70% of the population are farmers and they are the target group affected by the hazards of pesticide applications. Moreover, in developing countries farmers are rather careless in handling pesticides. All pesticides in a given chemical group generally affect the human body in the same way; however, severity of the effects varies depending on the formulation, concentration, toxicity and route of exposure of the pesticide. It is important, therefore, to know the type of pesticide associated with poisoning. Data collected so far reveals weight loss, anaemia and splenomegaly. There are several diseases linked with pesticide exposures, but in our clinical examination we found development of CML-like complications.

Organophosphorus (OP), organochlorine (OC), carbamate(CA) and pyrethroid pesticides are known to cause acute toxic effects in workers who are directly exposed to these lethal chemicals during spraying. Pregnant women who are exposed during spraying periods can also be affected and so also young children accompanying the adults. Indirect exposure here refers to the after-spray effects of vector control and repellents used for indoor pests. Pristine environs include those places which are away from pesticide treated fields, generally found in forests and sometimes in urban areas. Individuals exposed to pesticides who came to the hospital had health problems like headache, fatigue, dizziness, loss of appetite with nausea, stomach cramps and diarrhoea. Some had problems like excessive sweating and salivation while others complained of chest discomfort and tightness with involuntary urination and bowel movement and some had slower heartbeat; all these are the symptoms of pesticide poisoning.

The degree of toxicity depended upon the chemical nature of the pesticides and their concentrations. We have traced the effect of pesticides at molecular level at low doses. In almost all experiments we could determine chromosomal aberrations and DNA damage at $1/10^{th}$ of LC_{50} dosage of the pesticides like monocrotophos, chlorpyriphos, endosulfan, dimethoate, profenofos, and carbofuran. Chromosomal analysis of individuals exposed to pesticides showed few aberrations like gaps and breaks, and there was significant increase in the satellite associations compared to the normal individuals. It was found that the aberrations were more in those with longer exposure. No chromosomal abnormalities in any of the patients showing neurological or haematological disorders were observed. There were no signs of genetic predisposition, consanguinity and clear absence of Philadelphia chromosome. These studies suggest that pesticides have the capacity to alter the genetic material, particularly chromosomes in mammalian cultures.

The leukaemia-like condition observed here was not a result of chromosomal aberrations as indicated by the normal karyotypes of the subjects. Moreover, all the cases showed a positive response to other treatment regimens like Hydroxyurea, Methotrexate, Cytarabine, Prednisone and Vincristine and have a currently normal health status. There was no recurrence of the leukaemia-like condition in the individuals, thereby indicating that pesticide toxicity can be reversed by proper treatment.

Our observations also indicate that prenatal exposure to xenobiotics/pesticides might have led to mental retardation in 3% of the exposed children. Pedigree analyses of the children up to three generations revealed no genetic abnormalities and all the children showed normal karyotypes. Neurological complications and delayed milestones can be attributed to maternal exposure to pesticides during pregnancy.

Although these studies have been limited to 600 (400 adults + 200 children, both exposed and unexposed) individuals, it is noteworthy that many of the reported increased risks are of greater magnitude than those just observed, suggesting that Indians living in rural areas are exposed to the carcinogenic effects of pesticids to a much larger extent than urbanites. Future research on pesticides, especially in India, should include improved exposure assessment, evaluation of risk by the age at exposure, and investigation of possible gene-environment interactions. There is potential to prevent mental retardation and adulthood cancers by reducing or limiting exposure to pesticides.

A thorough examination of the sources of pesticide exposure suggests that avoiding consumption of foods grown with the use of pesticides is insufficient to protect a person from pesticide exposure. Individuals must also avoid or minimize cosmetic spraying of lawns and gardens and indoor spraying of pesticides to kill cockroach, flea, or ticks. Agricultural spraying of pesticides must be done with adequate warning to residents of nearby areas. Protection against pesticide poisoning is not an individual effort. A successful effort requires the cooperation of communities and the society as a whole. Therefore, adequate information about the potential risks of exposure and effects must be available for public examination in order to enhance regulations for pesticide usage. With adequate care and regulations, pesticides can serve as a useful tool in sustaining the world population instead of posing major health issues to society.

Therefore, unlike a previous notion that sticking to the stipulated limit of pesticide usage keeps us safe from its toxic effects, the present study indicates clearly that exposure to a much lower level can also prove to be harmful. The most effective way of handling the situation appears to involve a concerted effort of the scientific and regulatory bodies. The first step, however involves the thorough understanding and reanalyzing of the negative impacts of these chemicals. It is also necessary to devise mechanisms that avoid coming in contact with less toxic chemicals and avoid using the more toxic ones. A more promising alternative is the use of biological pesticides, the development of which is gaining momentum. Replacing the conventional pesticides with these alternative forms whereever possible will reduce toxic loads in biological systems and in the environment as a whole.

CONCLUSIONS

- The results indicated that environmental toxicants and xenobiotics were the cause of mental retardation in children, may be due to abnormal levels of acetylcholine in these cases, while hematological disorders were prevalent only in adults.
- Although these studies have been limited to 600 (400 adults +200 children, both exposed and unexposed) individuals, it is noteworthy that many of the reported increased risks are of greater magnitude than those just observed suggesting that Indians living in rural areas are more exposed to the carcinogenic effects of pesticides.
- Future research on pesticides especially in India should include improved exposure assessment, evaluation of risk by the age at exposure, and investigation of possible genetic-environment interactions.
- There is potential to prevent at least some mental retardation and adulthood cancers by reducing or limiting exposure to pesticides.

ACKNOWLEDGEMENTS

I am grateful to my student Ms. Prabhavathy Das who has contributed tremendously to this project, and I am thankful to the Directors of Mahavir Hospital and Indo American Cancer Hospital, Hyderabad for the support to this work and for the facilities provided.

References

Cook, J.A. & I.B. Mitchell. 1989. Viability measurements in mammalian cell system. *Anal. Biochem*. 179: 1-2.

Dearfield, K. L., H. F. Stack, J. A. Quest, R. J. Whiting & M. D. Waters. 1993. A survey of EPA/OPP and open literature data on selected pesticide chemicals tested for mutagenicity. *Mutat. Res.* 297:197-233.

Jamil, K., Abjal Pasha Shaik, M.Mahboob & D.Krishna. 2004. Effect of organophosphorus and organochlorine pesticides (monocrotophos, chlorpyriphos, dimethoate and endosulfan) on human lymphocytes in vitro. *Drug and Chem.Toxicol.* 27: 133-135.

Kaloyanova, F. 1985. *Hygienic Toxicology*. Sofia: Med.i Phizk

Moorhead, R. S. et al., 1960. Chromosome preparations of leucocytes cultured from human peripheral blood. *Exp. Cell Res.* 20: 613–616.

Olive, P.L., G.Frazer, J.P.Banath, C. Ferrarotto, J.R.N. Mclean & J.P Mcnamee. 1993. Radiation induced apoptosis measured in TK6 human lymphoblast cells using the comet assay. *Radiation Res.* 136:130-136.

Reddy P. J., D. Krishna, U. Suryanarayana Murthy & Kaiser Jamil. 1992. A microcomputer FORTRAN program for rapid determination of lethal concentrations of biocides in mosquito control. *CABIOS* 8: 209-213.

Yadav, J. S. & V. K. Kaushik. 1997. Genotoxic effects of ammonia exposure on workers in a fertilizer factory. *Indian J. Exp. Biol.* 35:101-104.

Zurlo, J., D.Raducille & G. Goldberg. 1996. The three Rs: the way forward, *Environ. Health Perspect*. 104:878-880.

3

Gender Inequity in Agriculture

SWARNA VEPA

ABSTRACT

The paper examines some important issues of gender inequity in Agriculture. The issues are women's contribution to agricultural GDP and their access to agricultural resources such as land, water, credit and technology. The paper also examines gender discrimination in agricultural wages. The paper is in three parts. The first part examines the contribution of women to agriculture in terms of time spent by women on activities and number of women engaged in productive agricultural activities with the help of the data from time use surveys and the National Sample Survey Organization's Reports and the 2001 census. The second part of the paper then analyses Women's poor access to land, water, credit and technology that reduces their productivity. The second part also elaborates how the women's work is undervalued in the economy resulting in a huge amount of unpaid work in family agricultural enterprises and low wages in paid work. The third part explicitly elaborates the adverse consequences of such imbalance in terms of misallocation of resources to unsustainable agricultural production as against sustainable production.

INTRODUCTION

It has been widely accepted that women are an integral part of the agricultural production systems in India and they contribute substantially to the agricultural income of the country. Agriculture including crop production, animal husbandry, fisheries, forestry, initial processing and storage is highly dependent on the time devoted by women. Agriculture is a low productivity sector. The share of agriculture in the national income has declined to 18.5% in 2006-07 as per the Economic Survey, but it continues to support close to 58% of the population. This means the per capita productivity in agriculture is very low. World's poverty is concentrated in agriculture and among the women in agriculture.[1] Within this low productivity sector, women get much less than men. Most of the work of women in agriculture is either unpaid or low paid. While women bear the burden of work they remain as helpers in household enterprises or agricultural labour, without any recognition. Unless women in agriculture are recognized for their work and receive their due share in the income and resource, higher growth of agriculture remains a distant dream. Shifting agriculture into the hands of women without any commensurate reward is neither good for agriculture nor the women who sustain it.

The present paper examines the issues of women's contribution to agricultural GDP and agricultural work and the limited access of women to income from agriculture and resources. The paper first examines the contribution of women to agriculture in terms of time spent by women on activities and number of women engaged in productive agricultural activities with the help of the data from Time Use Surveys and the National Sample Survey organization's Reports and the 2001 census. The paper also examines the extent of feminization of agriculture. The next issue taken up is the extent of unpaid work and un-recognized work of women and lower wages for women compared to men.

CONTRIBUTION OF WOMEN TO AGRICULTURE- ISSUES INVOLVED

It is now widely contended that the National Account Statistics underestimate the contribution of women to Gross Domestic Product (GDP) in general and agricultural GDP in particular. Contribution of women to agricultural GDP is rather difficult to estimate in India. The contribution depends upon two factors; the exact number of female workers in the agricultural labour force and the average value of their work.

1. *FAO and World Bank (2001),* Farming Systems and Poverty: Improving Farmers' Livelihoods in the Changing World. Edited by John Dixon and Aidan Gulliver with David Gibbon as Chief Editor: Malcolm Hall, Rome &Washington, DC

The major issue is the estimation of female workers and the work participation ratios. The contention is that the census as well as the national sample survey data underestimates the female workers in the work force. Work participation of women in productive work is much higher than what appears in National Sample Survey (NSS) data. The estimations of Time Use Surveys based on the principal and subsidiary status given in the published report of the Time Use Survey are not very different from those of National Sample Survey estimates of weekly status. The report only compares the rural and urban total worker participation rates with NSS 2000 data & puts the participation as being higher by about 6% for rural persons.[2] As per the principal and subsidiary status, the participation ratios remain at about 24.81% for females both in rural and urban areas put together as per the Time Use Survey and they are close to the ratio of the weighted average for urban rural areas for female workers in the National Sample Surveys. Thus obviously, if we go by the usual status of a person, which is determined by time criterion of having worked more than 180 days either by principal status or subsidiary status in a year, many female workers are considered as not being in the labour force.

The weekly status definition of a worker is expected to include more persons. According to this definition during a reference week if a person has worked even for one hour that person has to be considered as being in the labour force. Even by this definition, female worker population ratio remains only at 27.5 in rural India in 2004-05. The NSS 2004-05 worker population ratios are higher than in 1999-2000.[3]

However if this same criterion is used in the Time Use Survey[4] both male and female work participation ratios dramatically go up; Male worker participation ratio goes up to 60% and the female worker participation ratio goes up to 50%. Some authors with access to original data from Time Use Surveys compiled the information on the number of persons reporting for work during the reference week. The underestimation of NSS data was of the order of about 10% for males and about 25% for females.[5] There are wide variations among the six

2 Central Statistical Organization, Ministry of Statistics and programme Implementation, Government of India, "Report of the Time Use Survey" April 2000, New Delhi. Page 44

3 National Sample Survey Organization, "Employment and Unemployment Situation in India , 2004-05, "NSS Report No. 515. page 91

4 Unfortunately Time use pilot survey of 2000 covers only six states and the sample is much smaller than the NSS surveys. It is a one time survey and not repeated to study changes over period.

5 Indira Hirway (2002). Employment and unemployment situation in 1990s- How good are the NSS data?" *Economic and Political Weekly* (25 May)

states in worker population ratios of females. The underestimation for female workers was about 38.46% in Haryana, 32% in Orissa, 18.73 % in Madhya Pradesh, 12.15 % in Tamil Nadu and 8.67% in Gujarat in 1998-99 compared to the 1999-2000 NSS data. Unfortunately, the figures for rural agricultural workers separately for males and female are not made available from the published sources.[6]

An earlier study of Time Use estimated the work participation ratios for selected villages in West Bengal and Rajasthan in 1976-77 and compared with NSS 1977-78 (32nd Round) data for rural females. The study also found the ratios much higher for Time Use Surveys than the National Sample Survey and variations existed across the states. [7]

The variations in "the levels of underestimation by National Sample Survey and census is less in the states, where the female work styles are close to the male work styles and the underestimation is more in some states where women work styles are very different from that of men and women remain home bound. "The main cause of underestimation is the work styles of women with predominance of time allocated to housework. The work styles are characterised by intermittent participation, over a life cycle, over a day and a week. It shrinks or expands based on the season and the demand for work." [8]

This view is also supported by the recent survey of NSS that studied the women who were categorised as housewives, but engaged in economic activities. About 58.8% of those who are said to be not in labour force are engaged in activities that substantially contribute to the National Income. Women are involved in work related to agriculture, kitchen gardening, dairying, livestock, poultry and collection of food, fodder, fuel, water etc. All these women and their work is ignored in the National Income Accounting (Table 1).

Thus, it makes sense, to expect work participation and estimation of female workers to be of the same accuracy as that of males, where the female workers are engaged in the work on fields outside home,

---

6 The original source of calculations is quoted by Indira Hirway as R. N. Pandey, " Estimating Workforce Participation rates Using Time use Survey data and its comparison with usual labour force survey -Indian Experience. *National Sample Survey Organization*, New Delhi.

7 Devaki Jain (1996) "Valuation of Work and time as a measure". *Economic and Political Weekly*, Vol. XXXI No. 43, October 26, 1996

8 *Ibid*

whether they are paid wage labour or own account workers on the family farms. Hence, female workers in crop production may not pose much problem of under estimation. The problems of underestimation may occur in the work performed in the household premises such as kitchen gardening, storing, cleaning, dairying, livestock rearing, fishing and forestry and collection of food, fodder, fuel, water etc. Underestimation of these types of economic activity is more likely as people in general do not consider it as economic work.

The discrepancy also occurs because of exclusion of certain types of work in the National Sample Survey, which is a part of the 1993 United Nations System of National Accounting adopted by Time Use Surveys. National sample surveys do not consider all production for self-consumption. It only considers primary production for self-consumption. Processing of primary products for own consumption is not an economic activity for the National Sample Survey but counts as economic activity in Time Use Surveys.

Table 1: No. of women of age 15 years and above usually engaged in household duties (principal status) and also participating in specified activities per 1000 women usually engaged in household duties (All India)

S. No	Specified additional activities	Rural			Urban		
		With	**Without**	**All**	**With**	**Without**	**All**
1.	Maintenance of kitchen garden	164	129	135	80	31	33
2.	Work in household, Poultry etc.	543	276	319	219	37	44
3.	Either item 1 or 2	591	328	370	262	62	70
4.	Free collection of fish etc.	179	147	152	46	11	12
5.	Free collection of fire wood etc	534	370	396	164	50	55
6.	Any of item 1,2,4 or 5	817	540	584	355	99	110

Source: NSS Report No.465,"Participation of Indian women in household work and specified activities", 1999-2000

€ Note: SS = Subsidiary work Status

Strictly speaking, Time Use Surveys are not comparable to census data or the National Sample Survey data due to the variations in the very definition of work, coverage and sample size involved. All the same, it is possible to capture the level of underestimation involved.

Based on the contention of Indira Hirway (2002) and others, the underestimation of male workers population ratio appear to be about 10% and that of female workers population ratio appear to be about 25%. Correspondingly, the population not in the labour force would come down. As against the work participation rate of NSS 61st round data for 2004-05, at 27.5% for females and 52.4 for males,[9] The Time Use Survey rates of work participation computed by some authors are 52.5% for female and 62.4% for males.

If we apply these ratios to the projected agricultural population of males and females, as of March 2008, there would be 174.97 million female workers and 221.72 million male workers in agriculture. The proportion of female workers in the total agricultural workers would be 44.2%, much higher than the 39% given by the 2001 census data [10] (Table 2).

Table 2: Estimated Percentage of Female Agricultural Workers

	Projected for March 2008
Total population	1,147,677
Agricultural population	688,606
Male population	355,321
Female population	333,285
Male workers	221,720
Female workers	174,974
Total Agricultural Workers	396,695
% of male workers	55.89%
% of female workers	44.19%

Source: Based on Projected population census of India and the time use work participation ratios of 52.55 for females

9 Indira Hirway 2002.

10 The census of India projected Population for 2008 and 2001 ratios of agricultural population and male and female proportions are used for the estimation of total male and female population in agriculture.

The time use surveys show that women's total contribution to work is apparent not only in crop production, but also in the other activities that form part of the System of National Accounts (SNA). In all the six states given in table 2, women had spent as much time or much more time on non-crop production SNA activities as on crop production. The non-crop SNA activities in agriculture include livestock, fisheries, forestry, collection of food, fodder, fuel, processing, storage, and so on. Women had spent an average of 10.20 hours per week on crop production activities in the year 2000. This alone has translated into 39% per cent of female workers to total workers as cultivators and agricultural labour. An additional 10.37 hours a week, was spent on non-crop agricultural activities by women. The total time spent on agricultural activities was about 20.57 hours per week by women, which would mean either more women workers in agriculture than reported in the Census or more main workers spending more time on these activities.

Valuation of Work

The next important issue is about the value of work performed by women. Valuation of the work and recognition of the worker gives visibility to the worker and proper measures adopted to address the concerns and improve the agricultural production. Estimating the collective contribution of women to agriculture further strengthens the need for gender sensitive policy measures.

Time Use Surveys clearly show that most of the work performed by women is unpaid (Table 3). The data clearly show that about 50% of the work performed by women is unpaid in all the six states put together. Some states such as Haryana, Meghalaya, followed by Orissa seem to have large proportions of unpaid work ranging from 62-85%

Most of the work performed by women in agriculture on family enterprises is unpaid work. It is difficult to value this unpaid work. Agriculture is both subsistence sector as well as informal sector. Though agricultural production for self-consumption as well as for the market has been valued and included in the GDP, there is a large scope for underestimation of the female workers and their output. Unpaid work is not free. It has opportunity cost and it is an economic good that contributes to the income of the household.

Table 3. Weekly Average Time Spent on Crop Production and Other SNA Activities

	1	2	3	4	5	6	7
State	**Crop Prod uction**	**Non-Crop Agricu ltural Activities**	**All Agric-ultural Activities**	**Other Activ ities**	**All SNA Activ ities**	**% of Unpaid Work**	**% of Female Workers**
Gujarat	11.99	9.63	21.62	2.28	23.90	44.67	41.71
Haryana	3.14	18.04	21.18	2.26	23.49	85.99	40.46
Madhya Pradesh	13.56	6.51	20.07	2.55	22.62	52.40	43.77
Meghalaya	13.57	12.32	25.89	3.23	29.12	76.39	45.80
Orissa	6.61	10.15	16.76	2.27	19.03	69.44	35.70
Tamil Nadu	12.35	5.57	17.92	5.54	23.46	32.45	45.52
Average	10.20	10.37	20.57	3.02	23.60	60.22	43.00
Combined States/ All India	10.80	6.95	17.75	13.38	31.13	50.52	39.07

Note: Columns 1-6: Compiled from 'Time Use Survey' April 2000, Column 7: Census of India 2001 (percent of female workers to all farm workers)

Some authors had argued that the time spent itself should be measured as value, since more accurate valuation of unpaid work is not possible. The general perception is that work contribution of women to agriculture is much higher than that of men, though the Time Use Surveys in India do not substantiate this view. There could be some under estimation even in the Time Use Surveys as well.

Further, the micro studies and single peak season studies would show a very high participation ratio by females. The village surveys show higher rates of participation than the surveys covering larger geographical area and bigger sample. The Time Use Survey of 1998-99 is spread over the entire agricultural year. Based on micro studies, for one season, Food and Agricultural Organization found that women performed 90% of the work related to rice cultivation in East Asia.

The Time Use Survey of 2000 shows that on the average, women spend more time than men on activities related to animal husbandry in some states in which animal husbandry contributes substantially to the agricultural GDP, as in Gujarat and Haryana. In respect of activities such as collection of water, fodder, fuel, wild foods etc., women spend more time than men do in all the States (Tables 4 & 5).

Table 4. Average weekly time spent on activities related to Animal husbandry in rural areas (Hours/ week)

S.No	State	Women	Men
1.	Haryana	9.8	4.78
2.	Madhya pradesh	3.75	7.72
3.	Gujarath	5.93	4.31
4.	Orissa	3.15	3.35
5.	Tamil Nadu	2.42	3.69
6.	Meghalaya	1.22	3.92
	Combined	12.8	15.89

Source: Report of the Time Use Survey 2000

Table 5. Average weekly time spent on activities related to collection of food, fodder, fuel, water etc; (Hours/week)

S.No	State	Women	Men
1	Haryana	7.78	1.57
2	Madhya pradesh	1.52	0.91
3.	Gujarath	3.0	1.4
4.	Orissa	5.95	1.92
5.	Tamil Nadu	2.78	0.55
6.	Meghalaya	5.29	2.95
	Combined	3.52	1.18

Source: Report of the Time Use Survey 2000

Feminization of Agriculture

This brings us to the question of the extent of feminization of agriculture. The feminization of agriculture is not specific to India. It has been noticed all over Asia, South Asia, Southeast Asia and in parts of West Asia[11]. It has been pointed out that the Indian case is different from that of East Asia. The feminization of agriculture generally takes place due to out-migration of males from low paid agriculture to high-paid industry. In India, some economists believe that this has been induced by the 'casualization' of work, unprofitable crop production and distress migration. Migration has been noticed to other rural areas, to urban slums and to highly labour-exploitative sectors of the economy such as construction[12]. Indeed, many authors emphasized the feminization of Indian agriculture. There has been an increased involvement of women in agriculture at the turn of the century compared to the earlier period. It has been noticed in micro studies and in many field studies as well as district level studies[13].

To study the feminization of agriculture one needs data on two time-periods. National Sample Survey data indicates the percentage of female workers in agriculture increased from 38.8% in 1993-94 to 41.6 % in 2004-05, an increase of about 2.4%. Gender disaggregated agricultural workers data from Census are not comparable as the agricultural workers in 2001 include marginal workers, whereas they were excluded in the 1991 data, to note the trend over the past decade. This is because the 2001 data include marginal workers and the 1991 Census only gives main workers. However, for rural workers as a whole for all India, comparable information is available. The workforce composition of rural India (main workers) shows a 4% shift of rural main workforce in favour of women (Table 6).

Table 6. Composition of Rural Main Workers (persons in millions) (Census of India 2001)

	1991	Proportion	2001	Proportion
Female	80.43	0.32	111.46	0.36
Male	168.60	0.68	199.20	0.64
M and F	249.03	1.00	310.66	1.00

11 Food and Agricultural Organisation (FAO) Rome (2003), Sustainable Development Department, *"Asia's Women in Agriculture, Environment and Rural Production"*, Country papers on Sustainable Development Dimensions.

12 C. P. Chandrasekher and Jayati Ghosh (1999), *Indian Economic Reform Process and the Implications of South-East Asian Crisis*, Employment and Training Department, International Labour Office, Geneva (Mimeograph)

13 Anuradha Talawar & Swapan Ganguly (2001) Feminizationof Inda's Agriculture workorce, West Bengal Agricultural Workers Union, Calcutta. (Mimeograph)

Female main workers are replacing male main workers in the rural sector. The reported magnitude however has been very small. This is probably because of classifying women in family farms and family enterprises as marginal workers. The women marginal workers in rural India have increased from 24.43 million to 51.12 million. Addition of total female workers to the rural economy has been 31. 03 million compared to an increase of only 30.6 million of male workers, which shows that the importance of female workers in the economy is increasing (Table 7).

Table 7. Composition of Workers in India - Rural (Persons in millions) (Census of India, 2001)

	1991 Female	2001 Female	1991 Male	2001 Male	1991 Total	2001 Total
Total Workers	80.43	111.46	168.6	199.20	249.03	310.66
Main Workers	56.00	60.34	166.29	169.33	222.29	229.67
Marginal Workers	24.43	51.12	2.31	29.87	26.74	80.98
% of Main W. to Total W.	69.62	54.14	98.63	85.01	89.26	73.93
% of Marginal W. to Total W.	30.38	45.86	1.37	14.99	10.74	26.07

What is more obvious and substantiated by the data is the feminization of back ward agriculture. Percentage of female agricultural workers in the total agricultural work is more in some states (Appendix table 1). In backward and hilly regions, there are more women than men in agriculture. Arunachal Pradesh, Himachal Pradesh, Uttaranchal, Maharashtra have more women than men. There are more women cultivators in hill regions of Himachal Pradesh, Uttaranchal and Arunachal Pradesh. More than 50% of the agricultural labour are women in many states such as Andhra Pradesh, Karnataka, Gujarat and Maharashtra

The role of women in agriculture becomes clear, if we consider the district level data. Out of the total 582 districts for which data are available, about 46 per cent of the districts have more female labourers than men. In some districts more than 60 per cent of agricultural labourers are women. Only 8.70 per cent of the districts have more than 50 per cent women as cultivators (Table 8).

Table 8. District level percent cultivators and agricultural labourers

Percent	Cultivators		Agricultural labourers	
	No. of Districts	Percentage	No. of Districts	Percentage
< 30	254	43.64	94	16.15
30-40	140	24.05	76	13.06
40-50	137	23.54	143	24.57
50-60	36	6.19	202	34.71
60-70	14	2.41	63	10.82
> 70	1	0.17	4	0.69
Total	582	100.00	582	100.00

The situation is particularly noteworthy in the backward districts (Table 9). In the 150 districts that have been categorized by the Planning Commission as the most backward districts and in which first Food for work programme was introduced and latter the National Rural Employment Guarantee Scheme has been launched, female labourers are predominant. Census data are available for such 142 districts. In 83.7 per cent of these backward districts, female agricultural labour has more than 40 per cent share. In 87 backward districts (61 per cent of the backward districts), female agricultural labour constitute 50-70 per cent. Female cultivators also constitute more than 40 per cent of all cultivators in about 41 backward districts (29 per cent of the backward districts). Thus in the backward districts agriculture has more than the average share of women as cultivators and labourers (Tables 9).

Table 9: Distribution of districts as per % of female workers in 142 Food for Work districts

Percent	Cultivators		Agricultural Labourers	
	No. of Districts	Percentage	No. of Districts	Percentage
<30	51	35.9	8	5.6
30-40	50	35.2	15	10.6
40-50	34	23.9	32	22.5
50-60	5	3.5	57	40.1
60-70	2	1.4	27	19.0
>70	0	0.0	3	2.1
Total	142	100.0	142	100.0

Note: Data for 9 Districts are not available in the Census 2001

Wages are likely to be high in peak seasons in the rice growing irrigated areas of east and west Godavari districts of Andhra Pradesh. Women cultivators are fewer than men in the most prosperous districts of the state. In more difficult rain-fed terrains such as Vizianagaram, about 40 per cent of the cultivators are women. While the sex composition of cultivators and agricultural labourers differs from one state to the other, within the states, it appears that agriculture in more backward districts attracts a larger percentage of women than men, and at the same time, more prosperous districts have fewer women than men. The case of Tamil Nadu also gives credence to the view. In the prosperous irrigated district of Tanjavur, the percentage of women agricultural labourers to total agricultural labourers is only 44, while in the backward districts like Ramanathapuram as high as 58 per cent of agricultural labourers are women. In districts such as Namakkal, Pudukkotai, Tiruvannamalai and Perambalur, 55 to 60 per cent of agricultural labourers are women. It is not clear from the census data whether they are employed in rice cultivation or other commercial crops. The reason for higher employment of women as labourers appears to be the lower wages paid to women. Thus, women seem to be pushed to poorest districts of the state where the wages are low compared to prosperous districts. They get lower wages than men even in these backward districts.

In the case of Chhattisgarh, in all the 16 districts, the percentage of female agricultural labourers to total agricultural labourers varied between 53 percent and 66 percent. In the district of Dantewada it is highest at 66 per cent and lowest at 53 per cent in Raigarh. Similar situation exists in Gujarat also. The percentage of female agricultural labourers to the total agricultural labourers varies from 60 to 70 in the districts of Dohad, Panchmahal and Dangs. Similarly, it is worth noting that the hill districts of various states especially Himachal Pradesh and Uttaranchal recorded larger number of women cultivators than men cultivators. In Uttaranchal, in all the districts women outnumber men as cultivators: ranging from 68 per cent in Almora and 67 per cent in Garhwal to 55 per cent in Uttarkashi. In nine of the 13 districts of Uttaranchal more than 55 per cent of the cultivators are women. In eight districts they are more than 65 percent.

Gender gap in the wages of casual labour

The plight of women in casual labour appears to be even worse. Not only do they confront shrinking opportunities and low wages, but they also face discrimination. The wages of female casual labour are particularly low in some states. Madhya Pradesh, Orissa, and Andhra Pradesh and Karnataka seem to pay the lowest wages for women in absolute terms, which is between Rs.23/- and Rs.27/- in 1999-2000. In 2004-05, the situation has not changed much. The average female wage in rural Chhattisgarh and Maharashtra was about Rs.28/- and in Orissa it was about Rs.29/- and wages were around Rs.30/- in Andhra Pradesh and Madhya Pradesh. It has been reported that female wages could fall to the lowest levels of Rs.10/- in drought-hit areas such as Mahabubnagar of Andhra Pradesh, that records 48% poverty in the lean seasons[14] .

To indicate the wage inequality among the states, female wage was taken as a proportion of male wage. Ideally, it should be one, indicating no wage differentials. In 1999-2000, the ratio was low in three states at 0.51 in Tamil Nadu and 0.56 in Goa and Kerala. In 2005-06, five years later, situation substantially improved in Goa and the wage gap was 0.67. In Tamil Nadu, the gap remained almost the same at 0.52, while in Kerala the situation deteriorated to 0.48. Though the wage levels are not very low in absolute terms in these states, the gender gap is high due to higher wages for men. In the past five to six years the rise in men's wage was higher than the rise in women's wage.

The female wage in absolute monetary terms declined in Jammu & Kashmir and the male wage increased by about 30 percent, widening the gap in 2005-06. In Kerala the female wage increased by Rs.10/- while the male wage increased by Rs.33/-, widening the wage gap. In some of the north eastern States, raise in male wages was high, while female wages had a modest rise. For the country as a whole, gender inequity in daily wages for casual workers increased. The female wage as a ratio of male wage has declined to 63% in 2004-05, compared to 65% in 1999-2000. The gender gap in wages widened for casual labour in ten states out of 30 states for which data are available. The average daily wage for casual labour in rural areas in 2004-05 (NSS 61st Round) has been adjusted (deflated) with the Consumer price Index for agricultural labour to make it comparable with 1999-2000 (NSS 55th Round). The real wage for female casual labour was almost stagnant, widening the gender gap in real wages (Appendix tables 1&2). The real wage/day increased by just one rupee for female labour and by Rs.2.66/- for male labour over five years (Table 10).

Table 10: Gender disparity in wages

Year	Female wage/day (Rs.)	Male Wage/day (Rs.)
1999-2000	29.00	44.05
2004-05 (Adjusted)	30.20	47.50
Change	1.01	2.66

The states in which wages are high, the percentage of female labour is also relatively low. In states such as Punjab where the wages are high and wage differential is relatively low, female agricultural labour constitute only 25% of total agricultural labour. In Andhra Pradesh and Karnataka where the female wages are quite low at Rs.26/- and Rs.27/- in 1999-2000, the percentage of female labour in the total agricultural labour is high at 46% and 43% respectively. The absolute wages are the lowest for women in Madhya Pradesh and Orissa and they have large percentage of female labour to the total agricultural labour. About 48 percent of agricultural labour in Orissa and about 53 percent of agricultural labour in Madhya Pradesh are women.

Indian agriculture no doubt is sustained on the backs of rural women who have been poorly rewarded in terms of income, access to resources and decision-making. It is important now to recognize this and provide equal wages to the female labour and resources such as land, assets and credit and suitable share in the agricultural income to the unpaid women workers in the family farms. They need to be recognized as farmers and given their due share.

Appendix Table 1. Percentage of Female Workers and Male-Female Wage Differentials

S. No	State	Percentage of Female			Wages		
		Cultivators to Total (M+F) Cultivators	Agricultural. Labourersto Total (M+F) Ag. Labourers	Farm Workers to Total (M+F) Farm Worke	Female Rs.	Male Rs.	Index of Wage Equality (Female/ Male)
1	Andhra Pradesh	33.34	53.75	46.24	26.48	40.67	0.65
2	Arunachal Pradesh	51.20	45.99	50.89	42.73	67.09	0.64
3	Assam	28.92	34.17	30.27	35.55	48.82	0.73
4	Bihar	20.59	35.11	29.62	31.57	36.53	0.86
5	Jharkhand	37.93	48.06	42.20	--	--	--
6	Goa	48.43	55.99	51.53	46.99	83.20	0.56
7	Gujarat	32.49	52.22	41.71	34.43	43.91	0.78
8	Haryana	38.83	44.43	40.46	51.01	62.65	0.81
9	Himachal Pradesh	57.41	40.83	56.67	50.36	67.06	0.75
10	Jammu & Kashmir	36.35	22.09	34.52	66.07	77.04	0.86
11	Karnataka	29.61	58.66	43.20	27.14	42.51	0.64
12	Kerala	15.54	33.35	27.80	56.65	100.78	0.56
13	Madhya Pradesh	37.66	52.97	43.77	24.91	30.15	0.83
14	Chattisgarh	42.62	59.60	49.67	--	--	--
15	Maharashtra	44.12	56.73	50.16	25.28	41.32	0.61
16	Manipur	44.70	56.15	46.76	47.40	59.46	0.80
17	Meghalaya	45.19	47.45	45.80	43.06	57.37	0.75
18	Mizoram	49.28	52.48	49.44	66.24	97.77	0.68
19	Nagaland	50.07	46.19	49.85	46.67	71.93	0.65
20	Orissa	20.57	48.56	35.70	23.34	31.14	0.75
21	Punjab	13.37	25.90	18.52	49.48	65.86	0.75
22	Rajasthan	46.33	59.33	48.39	37.02	55.19	0.67
23	Sikkim	46.62	49.20	46.92	40.60	50.71	0.80
24	Tamil Nadu	35.93	51.52	45.52	30.78	60.20	0.51
25	Tripura	29.17	41.43	34.94	38.66	49.14	0.79
26	Uttar Pradesh	20.00	39.45	27.33	30.08	43.50	0.69
27	Uttaranchal	56.23	26.76	52.21	--	--	--
28	West Bengal	16.85	31.03	24.87	35.59	44.60	0.80
29	Chandigarh	6.60	8.14	6.93	--	--	--
30	Delhi	29.47	29.18	29.39	N.A	80.99	(-)
31	Pondicherry	10.80	43.28	39.15	--	--	--
32	Daman & Diu	50.60	75.58	56.79	--	--	--
34	Lakshadweep	0.00	0.00	0.00	--	--	--
35	A & N Island	30.01	22.19	28.52	--	--	--
	All India	32.59	46.91	39.07	29.01	44.84	0.65

Source: Columns 1-3: Census of India 2001; Columns 4-6 NSS 55th Round, 99-2000

Appendix table 2. Average daily wages (Rs. 0.00) for casual workers of age 15-59 years (Rural) 2004-2005

S.No	States	Male Wage	Female Wage	Wage Difference
1	Andhra Pradesh	50.30	30.88	0.61
2	Arunachal Pradesh	104.38	50.60	0.48
3	Assam	62.59	53.29	0.85
4	Bihar	45.41	37.42	0.82
5	Chhattisgarh	37.60	28.55	0.76
6	Gujarat	52.80	43.17	0.82
7	Haryana	75.26	59.34	0.79
8	Himachal Pradesh	88.88	62.00	0.70
9	Jammu & Kashmir	100.24	53.39	0.53
10	Jharkhand	51.11	37.93	0.74
11	Karnataka	48.33	30.74	0.64
12	Kerala	134.86	65.75	0.49
13	Madhya Pradesh	38.58	30.53	0.79
14	Maharashtra	47.37	28.16	0.59
15	Orissa	42.29	29.65	0.70
16	Punjab	75.14	53.10	0.71
17	Rajasthan	64.33	52.03	0.81
18	Tamil Nadu	70.45	36.53	0.52
19	Uttaranchal	69.66	56.20	0.81
20	Uttar Pradesh	53.37	39.54	0.74
21	West Bengal	49.88	30.88	0.80
	All India	55.03	34.94	0.63

Source: NSS Report No. 515-2004-2005

4

Status of Composting in India With Special Reference to Vermicompost

SULTAN AHMED ISMAIL, PRISCILLA JEBAKUMARI AND DAKSHAYANI GANESH

ABSTRACT

Several methods of composting are in practice in India such as the anaerobic pit composting, Indore method, NADEP method, bio dung method and vermicomposting. Wastes at the domestic, community and industrial level, can be converted to compost as a viable solution in solid waste management, as well as compost production as a viable micro-enterprise functioning as income generation program. Agricultural waste and residues are an excellent resource for composting at rural centres. Beneficial use of this compost in organic farming as well as in soil bioremediation has been well proven.

A complete succession of micro organisms comprising of bacteria, fungi, actinomycetes and algae as well as the change in physical and chemical parameters during the composting process (bio dung followed by vermicomposting) have been elaborated in this paper. The simplified version of vermicomposting, equipment needed, composting, bacterial degradation, harvesting benefits, economics and dosage of vermicompost to be applied for various types of cropping are dealt in.

INTRODUCTION

Solid waste management has emerged as one of the biggest challenges today in urban India. The municipal bodies have been unable to adjust to the rapid changes that have led to both increased quantities and changes in the composition of the waste. The current situation, which gives rise to the indiscriminate dumping and burning of wastes, has a serious impact on air, land and water pollution and causes a dramatic increase in health hazards in the urban environment.

As a response, in many cities non-governmental voluntary organisations (NGOs) have started developing neighbourhood waste collection services, as well as initiating composting and recycling activities. These moves are backed up by new policies of municipal solid waste management aided by handling rules published in 2000, which, among other recommendations, require source segregation and waste recovery.

Several alternatives have been proposed for solid waste management by composting. Some recommend incineration. However, recycling the materials saves three to five times the amount of energy as incinerating these same materials would require.Waste incineration encourages a one-way flow of materials on a finite planet, thus making the task of conserving resources and reducing waste more difficult, not easier.

Today the world is moving towards Zero Waste Management which aims to recover at least 95% of the total recyclable fraction of household waste so that only 5% need to be handled by the Municipal Corporation. There are cases where countries are starting to adopt Zero Waste Goals as policy measures. San Francisco has already achieved 52% landfill diversion and is hoping to reach zero status by 2020. In India, the Supreme Court (MSW Rules 2000) has issued a directive to implement the same before 31st December 2003. Sulurpet in Andhra Pradesh is one of the few municipalities which has implemented source segregation, door to door collection and composting. Recently Namakkal in Tamilnadu has achieved the same.

The best way to handle most of the household organic waste is to compost it. Composting is a controlled oxidative process that involves a heterogeneous organic substrate in the solid state, evolved by passing through a thermophilic phase leading to the production of CO_2, water, minerals and stabilized organic matter called compost (Diaz *et al.*, 2002). Several methods of composting are in vogue in India and any one or many methods of composting can be applied in given situations. Selecting an appropriate method of composting appropriate to the place, based on resource availability, as well as limitations should be applied to evolve the most suitable model for a given place.

METHODOLOGY

Composting can be broadly classified into two types: the anaerobic and aerobic models. The anaerobic model was the most prevalent in the past in all rural centres, when all litter and organic refuse was put into pits and closed. The manure would be ready in six months, ready to be used as nutrient to the soil.

Today with changing cropping patterns and the need for quick availability of manure becoming invitable, many are resorting to aerobic model of composting. Biodegradation of natural materials will produce valuable compost as the major product, along with water and carbon dioxide. The CO_2 produced does not contribute to an increase in greenhouse gases because it is already part of the biological carbon cycle.

In rural sectors, the *NADEP Composting* has come into practice. This involves construction of tanks specifically designed for NADEP with air vents on the sides of the tank. The most convenient prescribed size for a NADEP tank is 12 feet (L) X 5 feet (W) X 3 feet (H). Each tank can take in 1,350 to 1,400 kg of dry and green waste combination. This needs to be supplemented with about 100 kg of cattle dung, about 1675 kg of fine sieved soil and about 1500 litres of water. The top of the tank is sealed with soil and preferably a shed is put on the tank. About 2.5 tons of compost is obtained from each tank in about 90 to 120 days. Though this is indeed a good practice, Windrow method of composting has taken over the NADEP method to save costs and space.

For municipal solid waste, the *Windrow Method* is very suitable. Shredding of the waste is essential to increase surface area. This shredded waste should be placed into windrows of any convenient length but with width not exceeding 4 to 5 feet and the height not exceeding 4 to 5 feet. For effective composting care should be taken to see there is air space within the pile, that the moisture content is at least 60% and the C:N ratio is less than 40. To make Windrow composting more effective, it is better to adopt the biodung method of composting.

The Biodung Method has the advantage of an aeration chamber at the base made with twigs, which enables quick heating of the pile or the windrow. Covering of the windrow with a polythene sheet enables moisture retention and build up of high temperature inside the pile, facilitating the destruction of unwanted seeds as well as any pests or parasites in the waste used. The windrow needs an occasional turnover at specific periods. The manure is ready in about 45 days to 60 days.

However those desirous of producing vermicompost can transfer the biodung material after 30 days, cool it enough to be acted upon by earthworms to produce vermicompost.

One of the important methods of composting is *vermitech*–that is involving earthworms in the composting process. Earthworm diversity necessiates species specific methods of culturing them during vermicomposting. Solid organic waste management using earthworms is being practiced in several methods depending on the species of earthworm. The procedure for handling small quantities of waste generated from households (indoor composting) is slightly little different from that applied in handling larger quantities of biomass from markets and agriculture (outdoor composting). Vermicomposting could be defined as "The combination of biological processes, design and techniques to systematically and intensively culture large quantities of one or more species of earthworms to speed up stabilization of organic waste materials, which are eaten, ground and digested with the help of other micro organisms like bacteria, fungi and other smaller animals". Organic matter then naturally gets converted into much finer particles like castings (faecal pellets from the earthworms). This compost is active microbially and important plant nutrients in them are in an available form to plants.

VERMICOMPOSTING

1. *Vermicomposting using local varieties of earthworms:* In India, so far five hundred and nine species of earthworms have been identified and thus it is indeed easy to collect the epigeics and anecics from garden soils after monsoon spells. These earthworms can be cultured or used in composting applying simple procedures in pits, crates, tanks, concrete rings or any containers (Ismail, 1994, 1997, 2005). Only after pre-processing or pre-digestion through partial anaerobic phase (done under black polythene cover or with a clay seal layer), organic material is ready for composting. The biodung composting technology is highly recommended by us as a pre-digestion mechanism. Local species of earthworms that are generally used in vermitech (vermes=

earthworms; tech= technology) in India are *Perionyx excavatus* and *Lampito mauritii*.

2. *Vermicomposting using exotic species of earthworms*: Exotic species of earthworms have also been used in India for vermicomposting. Internationally three species of earthworms have been found ideal for vermicomposting namely, *Eisenia fetida* and *Eudrilus eugeniae*, which are exotic, and *Perionyx excavatus*, which is endemic.

Microorganisms in the process of composting (Fig.1): The process of composting, involves different microorganisms such as bacteria, fungi, actinomycetes, phosphate solubilizers and other microorganisms involved in the Nitrogen cycle. Succession is observed in the quantity of microbes depending upon the nature of the substrate, the age of the compost, the ambience created by the existing microbes to its successors and also the physical and chemical characteristic of organic material used. There also occurs diversity within each group of microorganisms. In the microbial succession during the composting process, decomposers are succeeded by the biofertilizers.

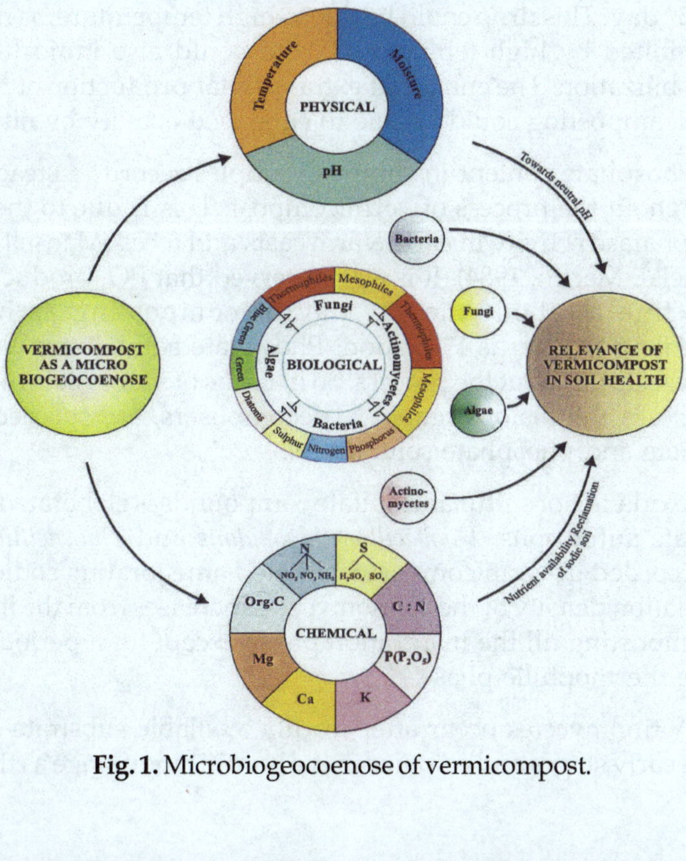

Fig. 1. Microbiogeocoenose of vermicompost.

The majority of the microorganisms in the initial stages of the composting are the heterotrophic bacteria, which rely on the oxidation of the large amount of organic Carbon. They get reduced during the thermophilic phase until the formation of the biodung compost. They then increase during the passage of organic material through the earthworm gut in the presence of the assimilable C in the gut and subsequently in the cast of the earthworms (Lavelle & Martin, 1992).

There is increased presence of ammonifiers in the initial stage of composting, which correlates with the high amount of protein degradation and contributes to reduction in C:N. Nitrifiers however increase from the initial to the final stages. The products of the ammonifers create an environment for the multiplication of nitrifiers which utilize ammonia and convert it to nitrite and nitrate.

Nitrification potential is indicated by NO_2-N decreases with composting time. The NO_2 production drops and stabilizes to low levels during the later stages of composting till no further decomposition can take place, as the C: N ratio gets stabilized. The NO_3 production increases till about the 14th day of composting thereafter declining till the 35th day. This drop could be due to high temperature, as nitrification is inhibited by high temperature and could also indicate microbial immobilization. The enhanced extra-cellular production of NO_3 during vermicomposting could be due to enhanced activity by nitrifiers.

Phosphate content in compost samples records a steady increase throughout the process of vermicompost. This is due to the increased phosphatase activity in earthworm casts and feces (Mansell *et al.*, 1981; Satchel & Martin, 1984). It is also observed that PO_4 production shows a decline at about the 21st day of composting as high NH_4^+ concentration retards P fixation. Phosphate solubilizers also steadily increase throughout the process. So in terms of succession, ammonifiers which are the major organic N decomposers, are succeeded by the nitrifiers and phosphate solubilizers.

Oxidation of sulfur and sulfate compounds is elaborated by aerobic obligate autotrophs. *Thiobacillus thiooxidans* and *Thiobacillus thioparus* are recorded in vermicompost facilitated ameliorating sodic soils. The population density of the actinomycetes increases from the initial phase of composting till the maturation phase except for a period of decline in the thermophilic phase.

Actinomycetes occur after readily available substrate disappears in the early stages and colonize in the humification stage as the compost reaches maturity. The optimum temperature for actinomycete activity is 40-50° C, which is also the temperature for lignin degradation in compost (Tuomela *et al.,* 2000).

Fungal density decreases as the composting process progresses. Mucoraceous group of fungi commonly referred to as sugar fungi are observed in the initial and early phases of composting which is in agreement with the nutritional hypothesis (Webster, 1970). Species of *Aspergillus dominate* and are responsible for major degradation of initial organic carbon as they are known to elaborate cellullases and hemicellulases. A lignolytic fungi, *Coprinus* spp. is predominantly found to colonise the compost towards the end when complex organic matter is biodegraded.

The thermophilic fungi record an increase in density and diversity during the thermophilic phase and bring about degradation of cellulose, lignin and pectin at a faster rate in conjunction with high temperature. *Trichoderma viridae* and *Trichoderma harzianum*, both potential biocontrol agents are present during the composting process and to a greater extent in the vermicompost.

The density and diversity of algae increases progressively reaching maximum density in vermicompost. Of special significance are the presences of algae such as *Oscillatoria* spp. *Anabaenpa* spp. and *Nostoc* spp. which are known to enhance soil fertility.

TECHNOLOGY OF VERMICOMPOST

To simplify the loading procedure for composting in Institutions and in the rural sectors where the availability of organic material is in bulk, a four-tank system has been developed based on a combination of biodung and vermitech technique that enables continuous compost production using cattle dung, weeds, leaf litter and other farm waste. In the same manner two-tank system is also suggested for lesser quantities of garbage.

HARVESTING OF COMPOST

When the compost is ready for harvesting, moistening of the pit should be suspended for 3 to 4 days prior to harvesting. Spread the compost after harvesting, collect the worms, if any for further inoculation, air dry the compost and sieve through a 2.5-3 mm sieve. The material is then ready for use or to be packed for later use.

Table 1. Four Tank Loading Schedule

1.Days	Tank	Process
00 - 30	01	Collection of biomass and cattle dung
30 - 60	01	Addition of cattle dung, water, covering with black polythene (Biodung)
	02	Collection of biomass
60 - 90	01	Inoculation of earthworms
	02	Biodung preparation
	03	Biomass collection
90 - 120	01	Vermicompost ready ; earthworms migrate from pit 1 to pit 2
	02	Vermicomposting
	03	Biodung preparation
	04	Biomass collection
120 -140	01	Harvesting of compost and collection of biomass
	02	Vermicompost ready; earthworms migrate from pit 2 to pit 3
	03	Vermicomposting
	04	Biodung preparation

BENEFITS

Composting is the best method of garbage handling given our temperature and environment. The type of composting applied shall depend on the given situation and area. Biodung composting or Windrow composting are the most preferred method for SWM. However due to the additional qualitative value of vermicompost, vermitech can be applied wherever necessary.

ECONOMICS

Whatever may be the technology applied, the selling price of manure in the market has a wide range from Rs.5 (~ 0.1 $) a Kg to Rs.20 (~ 0.4 $) a Kg depending on the type of outlet. The quality of manure is also equally important.

Investment costs vary according to the space, land, and water and resource availability. A simple unit handling one ton of waste per day through biodung Windrow method with suitable land, water and free access to resource material (organic biomass) may cost about Rs. one lakh (~ 2000 $) only. Generally 20 to 25% of one ton organic waste gets converted into manure.

EQUIPMENT

The need for mechanisation increases with higher production of compost. For 1 to 2 tons of waste per day manual handling is sufficient. For large quantities, transport equipment, front-end loaders, pulverisers/ shredders, windrow turners, screeners, packing equipment etc may be needed. An area about 5000 sq ft shall suffice to handle one ton per day of garbage.

RECOMMENDED DOSAGE

Compost is essential for soils as organic matter plays a vital role in crop production. A general recommendation is given below; however suitable changes should be adopted as and when necessary depending on crop and the type of soil.

100 gm of compost is applied for a pot containing 8 to 10 kg soil.

1-10 kg of compost per tree, depending on age and size of tree.

4 to 5 tons of compost per hectare of land.

Regular watering and mulching of the land is important.

No chemicals should be sprayed over the compost pit.

Sprays of plant origin are recommended, only if necessary, for plants.

OTHER APPLICATIONS

Many industries produce two streams of waste water from their treatment units - the sewage treatment plant (STP) and the effluent treatment plant (ETP). Sludge from the STP has been handled very successfully with appropriate biomass and converted into vermicompost. Sludge from the ETP if free from heavy metals has also been handled with suitable bulking materials through modified biodung composting technologies.

CONCLUSIONS

Composting is a process in which most of the nutrients are mineralized; the remaining complex substrates such as lignocelluloses could also help in the formation of humus. Moreover the microbial population in the compost continues its activities in the soil on application, acting as a "factory" enabling the organic decomposition process. The propagation of modern concepts where dried and powdered organic material is recommended for application to soils may initiate a thermophilic phase in the soils damaging tender roots. Care has to be taken that appropriate composting procedures are applied to prepare manure from organic wastes.

ACKNOWLEDGEMENTS

We thank the managements of The New College, Dr MGR Janaki College and the Stella Maris College, for support and facilities. Thanks to a large number of organizations and individuals with whom I associate myself in the composting process.

References

Diaz, M. J., E. Madejon, F. Lopez, R. Lopez & F. Cabrera. 2002. Optimisation of the rate vinasse/grape marc for co-composting process. *Biochemistry* 37: 1143-1150.

Ismail, S. A. 1994. Vermiculture in India. *Organic matters* 17: 23-28.

Ismail, S. A. 1997. *Vermicology: The Biology of earthworms.* Orient Longman, India. 92pp.

Ismail, S. A. 2005. *The Earthworm Book.* Other India Press, Mapusa, Goa. India. 101 pp.

Lavelle, P. & A. Martin. 1992. Small-scale and large-scale effect of endogeic earthworms on soil organic matter dynamics, soils of the humid tropics. *Soil Biol. Biochem.* 24: 1491-1498.

Mansell, G. P., J. K. Syers & P. E. H. Gregg. 1981. Plant availability of phosphorus in dead herbage ingested by surface casting earthworms. *Soil Biol. Biochem.* 32 (3-4): 163-167.

Satchell, J. E. & K. Martin.1984. Phosphatase activity in earthworm faeces. *Soil Biol. Biochem.* 16(2): 191-194.

Tuomela, M., M. Vikman, A. Halakka & M. Itavaara. 2000. Biodegradation of lignin in a compost environment: a review. *Bioresource Technology* 72(2): 169-183.

Webster, J. 1970. Coprophilous fungi. *Transactions of British Mycological Society* 54: 161-180.

5

Technological Empowerment of Women in Agriculture

MUSHTARI BEGUM, J

ABSTRACT

Nutrition is basic to sustainable development being concerned with ensuring food at all times in terms of quality and quantity. As a bridge between food and health it interlinks several variables such as environment and protection from illness and disease, which helps to lead to an optimal quality life. Despite the substantial progress made in India's food production, serious nutritional challenges continue to threaten her development in human resource development more so rural women are particularly vulnerable to malnutrition. Poor nutritional status of women during adolescence and their resultant vulnerability to anaemia during pregnancy contributes to a higher level of low birth weight children and if it happens to be girl child then the vicious cycle continues. Although food is available through markets at the district and village level, food security at the house hold level becomes a challenge because of lack of purchasing power. Women contribute significantly to agriculture production; on an average their contribution will be around 55-65 per cent. Hence, conscious effort is needed to develop an intervention strategy. One such strategy would be empowerment of women through transfer of technology, which will help in income generation especially during lean agriculture seasons. Priority areas in the field of nutrition for technology transfer would be processing, preservation and value addition of agro based crops. However, training is required for skill upgrading and capacity building of women.

6

Entrepreneurial Opportunities for Women in Peri-urban Horticulture

LEELA SAHIJRAM

ABSTRACT

Peri-urban horticulture is fast catching on as a viable business proposition. Participation of women is envisaged as a significant input in such an enterprise. Tissue culture propagation or ''Micro propagation' is a good candidate in this context among several biotechnologies available today. Micro propagation as a means of rapidly multiplying plants in a relatively small space irrespective of the season of the year has steadily gained popularity. Planting material in horticultural crops like banana and several other vegetatively propagated crops like ornamentals is in perpetual demand and can be met economically if women launched micro propagation as a peri-urban cottage industry. As against low rates of multiplication with conventional technologies, it is possible to raise thousands of propagules from a single starter-culture in a year's time using newer technologies. Details of setting up the industry on a home-scale and participation of women in peri-urban areas in this enterprise are discussed in this plenary talk.

INTRODUCTION

It is possible to demystify some of the 'elitist' technologies and bring them into the realm of the farm-level woman worker so that she can actively participate in and meaningfully contribute to national development besides contributing her mite as a home-maker. Tissue culture propagation or ''Micro propagation' is a good candidate among the several biotechnologies available today. Micro propagation as a means of rapidly multiplying plants in a relatively small space irrespective of the season of the year is steadily gaining popularity. Planting material in horticultural crops like banana and several ornamental crops is in perpetual demand and can be met economically if women launched micro propagation as a cottage industry. As against traditional low rates of multiplication using conventional technologies, it is possible to raise thousands of propagules from a single starter culture in a year's time using the newer technologies. Several ornamental crops have been micro propagated successfully at the cottage industry level as demonstrated by housewives in Kerala with orchids and anthuriums.

Micro propagation is a means of rapidly multiplying plants in a relatively small space under environmentally controlled and aseptic conditions independent of the season of the year. Millions of plants can be produced at any given time using this technology that has no parallel in any of the traditional propagation methodologies. Tissue culture has been applied successfully all over the globe but the emphasis has been on establishing high-volume, sophisticated commercial laboratories that turn out micro plants on a mega scale. This paper aims to conceptualise low-cost, large-scale micro propagation systems for producing high-quality plants for widespread agricultural use with women as the chief players. In this context, an *entrepreneur* is defined as a Person who organizes and manages a commercial undertaking" and *peri-urban* is defined as "Located in close periphery or close proximity of cities".

About the Tissue Culture Technology: In plants that are propagated asexually and in crops that need to be planted year after year, e.g. banana, this technology is a boon as it affords quick production of identical clones of the source/mother plant in huge numbers. There is a perpetual and heavy demand for banana propagule material in the country. Traditionally, banana is multiplied using daughter suckers that arise at the base of the pseudo stem. The farmer usually depends on neighbourhood farms for supply of suckers for raising a crop. This imposes certain restrictions, for instance, the desired variety may not be available at the time of need or in the quantities required. Besides, the farmer runs the risk of ending up with infected suckers that may result in yield losses up to 100%. Such constraints can be overcome by resorting to tissue culture method of propagation or 'Micro propagation'. The advantages of micro propagation are:

1. Large number of uniform propagules can be generated in a relatively short period of time with predictable dates of harvest.
2. The technology is season-independent whereby planting material can be produced round the year.
3. Variability encountered in size and propagule density (especially in clones suckering erratically) can be minimized since plants are grown under controlled conditions in vitro.
4. Allows for rapid bulking of novel clones when used in concert with breeding programmes.
5. Facilitates trans-continental exchange of disease-diagnosed planting material obviating quarantine hassles; far less air cargo space is required for transportation.
6. In vitro germplasm conservation is possible.

The national requirement for planting material of vegetatively multiplied seasonal crop species (sugarcane, potato, banana, ginger, turmeric, sweet potato, tapioca, garlic etc.) is estimated to be annually Rs.3000 crores. Although tissue culture is the ideal means of clonal multiplication, it has failed to achieve commercial success in this category of plants because of high cost of the *ex-vitro* plantlets (Rs. 6 to 10 per plantlet). ARTI (Appropriate Rural Technology Institute, Pune) uses tissue culture only to produce disease-free propagates in a nursery. Such planting material can be offered at prices acceptable to farmers. This concept was successfully introduced into Maharashtra in the case of sugarcane. Other candidate species for this technology are ginger, turmeric, banana, tapioca, elephant's foot, sweet potato, strawberry, etc., with banana ranking as the number one fruit crop produced by tissue culture. Thus, buying mother plants from a tissue culture laboratory and selling their nursery-grown progeny to farmers offers rural entrepreneurs business opportunity having a potential annual turnover of Rs. 3000 crores.

Planting material of banana is in perpetual demand and can be met economically if women launched banana micro propagation as a cottage industry. As against the traditional rate of multiplication (4-5x) obtained using suckers as propagules, it is possible to raise a hectare of banana plantation (i.e., 5000 propagules) from a single starter culture in a year's time.

Commercial tissue culture laboratories in India are into production of banana micro plants in a big way but the cost of production is prohibitive. Besides, these laboratories are located mainly in the cities and transportation to the end-users' farm adds to the cost of this perishable product. If a District-level or *Taluk*-level machinery could be evolved involving women entrepreneurs with low-cost inputs, the above constraints can be overcome and planting material can be supplied at cheaper rates.

Women as Entrepreneurs: The Government of India declared 2001 as the Year for Women's Empowerment. Numerous surveys carried out in India from time to time have all hammered home the point that women's co-operatives are run far more efficiently than those run by men. That most women do not harbour higher political ambitions is cited as a major reason for the success in running their co-operatives efficiently (Deccan Herald, Nov 12, 2000). The vibrant Indian dairy sector is a case in point in this direction, where, the small individual contribution made by ordinary, rural women folk has played a major role in making the White Revolution or Milk Revolution a spectacular success. Her participation as a co-operative member has, in addition, given the woman a modicum of economic independence and empowerment. As of now, 2 million women are active participants in the ongoing milk revolution encompassing 7000 village-level milk cooperatives in India.

This self-same example needs to be translated into the horticulture sector. The present concept paper is precisely aimed at such a possibility. If applied, this would result in income generation for the urban/peri-urban/semi-rural and rural women through gainful employment which is one of the themes of the session. Already an all-woman organization in Kerala, 'Nattika Vanitha Pushpa Krishi Samrakshana Samithi', Nattika, Thrissur–680 566, is engaged in production of orchids and other ornamental annuals through tissue culture. This example can serve as a shining beacon for women entrepreneurs intending to embark on a similar enterprise. In horticulture-based nurseries, women are traditionally the predominant workforce. This can serve as an additional advantage facilitating ready availability of skilled/semi-skilled labour.

Tissue Culture Enterprise on a Home-scale

A typical, family-operated nursery producing planting material in about 200 ha would have an annual turnover of about Rs. 20 lakh and an annual profit of about Rs. 5 lakh. Rural India would require 15,000 such nurseries to satisfy the total demand of planting material of this category of crop species, which can be met by tissue culture in a far lesser area and time.

Plant Tissue Culture (PTC) is an exciting area of biotechnology that can be used effectively and beneficially for improved yields. It is no longer the 'elitist' enterprise it was once considered to be. The need for extensive equipment, such as the autoclave and laminar air-flow hood, precision balance etc. has often limited the use of this technique. Of late, these equipments are not essential with the advent of the biocide PPM (plant preservation mixture) in the PTC media and preparation of media with cooker/microwave oven. PTC can be taken up at home with little problem of contamination. A description of these methods and reports of successful completion is given in *"Successful plant tissue culture in the classroom or home" by C.M. Stiff, Kitchen culture kits, Inc., Moscow, Idaho and plant cell technology, Inc., Washington D.C. E-Mail.KcK@turbonet.com*

Premixed powders for plant tissue culture are marketed in India by Hi-Media Ltd., Mumbai, but these mixtures do not guarantee freedom from contamination. There is a need to indigenize technologies to work on a home-scale for individual enterprise. One such alternate mode of setting up a successful outfit requires the following:

Microwave oven to sterilize media. Make sure the type of plastic is microwave compatible. Do a test run with a paper in it.

Pressure cooker to sterilize microwave-sensitive plastic lab ware/ metal dissection instruments.

PPM (Plant Protection Mixture), a biocide for minimizing contamination.

Simple fabricated box to prevent dust falling into culture vessels.

Premixed plant media to avoid purchase of individual chemicals and for preparation of stock-solutions (home-made media may suffice in particular cases).

Clean area: A simple plastic box can serve the purpose.

Bleach, detergent, vinegar, antacid, sugar, plates, knives, p^H strip, baking soda, etc. are available off a medical store or the kitchen-shelf.

Used jam-jars with autoclaveable caps/food containers can serve as culture vessels. These can be bought in bulk at a cheap rate from the local scrap-dealer.

Tap water/mineral water for media preparation can be microwaved for 10 minutes or processed in a pressure cooker for 20-30 minutes

Considering the vast potential of this business to generate self-employment in rural areas, ARTI, Puna has established a training-cum-production facility comprising a low-cost tissue culture laboratory and a nursery. The hardware cost of the tissue culture laboratory, producing about 20,000 to 50,000 plantlets per annum, would be only about Rs.2 to 2.5 lakh. The major cost-saving is achieved by using air coolers instead of air conditioners, pressure cookers instead of autoclaves and ordinary jam jars instead of "Pyrex" or "Borosil" ware. Also the running cost is drastically reduced by using rainwater instead of distilled water, and commercial grade sugar instead of Analar Grade sugar. The running cost of such a laboratory comes to hardly Rs.25,000 per annum. ARTI plans to annually train 15 to 20 persons in the procedure of micro propagation of sugarcane, banana, ginger and turmeric, hardening of the plantlets and secondary multiplication under nursery/field conditions. The facility would be operated as a commercial unit so that the trainees get exposure to the commercial aspect of the technology (Arve, 2000).

Here, we shall consider banana micropropagation as an example.

Socio-Scientific Paradigm of the Concept: In a mission mode, the objective of this exercise is to sustainably increase the productivity of banana and plantain grown on small holdings for domestic consumption and for local markets. The concept can fulfil the following socio-scientific objectives:

- To organize and co-ordinate a national effort on production of bananas involving woman power with an aim of quick dissemination of popular/improved cultivars.
- To promote and strengthen collaboration and partnerships in banana related industry at the regional/national level.
- To strengthen the ability of women to become self-reliant while participating in national development by using state-of-the-art Science and Technology (S&T)

- To co-ordinate, facilitate and support banana propagule production and distribution economically and to exchange information at a cottage industry level

Using the above approach, the focal theme of this symposium is addressed, which aims to:

- Increase women's access to land, resource and other capital investment
- Ensure adequate involvement in human resources and extension training programmes
- Increase the ability of women to generate income
- Make available the new tools of Science & Technology
- Target some of the new innovations and technology demonstrations to cater to women's needs
- Protect women's health and nutritional status
- Frame appropriate policies with the involvement of women

The Peri-urban Tissue Culture enterprise has several advantages namely Proximity to the city, Cheaper land prices than in the city, Access to raw material, excise items, etc., Proximity to airport wherever air-lifting is required, Access to farm women, Easy availability of *ex vitro* (soil) hardening medium and Tissue Culture Industry.

There are disadvantages too. These include quality of natural resources (land, irrigation, air quality), high population density; high cost of land, Leafy/root vegetables vulnerable to pollution effects, fruit crops demand large area and use of labour twice compared to cereals (456 days/ha in horticulture vs. 223 days/ha in paddy).

Modus operandi: There are two levels at which the technology can be made operative

1. Retrofitting tissue culture operations to involve urban/peri-urban/semi-rural/rural Women entrepreneurs
2. Individual /partnership level involving urban/peri-urban women entrepreneurs

(1) *Retrofitting tissue culture operations to involve urban/per-urban/semi-rural/rural women entrepreneurs:* Tissue culture procedures involving media preparation, clean area operations (inoculation

and subculture), primary hardening and secondary hardening can be carried out collectively by literate, semi-literate and illiterate women functionaries at different tiers of the technology.

Phase I: District headquarters can serve as the hub of the enterprise vested with the responsibility of exclusively carrying out in vitro operations. Specialized equipment required for such operations shall be housed at this laboratory and procedures involving plant collection, treatment, inoculation, incubation, subculture and prehardening would be the preserve of this tier of functionaries.

Phase II: Once the prehardened plants are ready for transplantation, these shall be transported to functionaries at the taluk headquarters/town where primary hardening operations will be carried out. This function can be performed by illiterate women functionaries too. The infrastructure required at this level is not as expensive as in the first phase of operations.

Phase III: Micro plants that have gone through primary hardening at the taluk headquarters/town need to be further transported to the village for secondary hardening. This facilitates easy supply of healthy and fully hardened micro plants to the end-user viz., the farmer. Secondary hardening entails use of huge quantities of good soil, sand and farm yard manure (FYM), which are readily available in rural areas. Skilled/semi-skilled rural women functionaries can operate at this level of the technology.

(2) *Individual/partnership-level involving urban/peri-urban women entrepreneurs :* The same scheme can be operated by an individual or by partners on a smaller scale under a single roof.

Entrepreneurial Opportunities for Women: Several national/state level organizations can be roped in as participants in development through such projects. Outfits like the National Horticultural Board (NHB) promote not only the development but also transfer of such technologies. For instance, the NHB provides 100% financial assistance up to Rs.10 lakhs per project for introduction of new technologies. (Proforma for seeking grant-in-aid from the NHB is available with the author). Under the horticulture promotional service, the NHB also encourages funding of projects relating to

- Improvement of nutritional status of rural masses
- Increase in per capita availability of fruits

- Help maintain ecological balance
- Utilization of land around homestead

The NHB has recognized tissue culture propagation of banana as a modern, transferable technology. Among organizations eligible for funding by the NHB are:

- NGOs
- Associations of growers
- Individuals
- Partnerships

Under development of commercial horticulture through production, projects related to production find favour for funding by NHB when they involve high-quality commercial horticultural crops, Biotechnology and tissue culture.

Banana fits the bill under the above components. In collaboration with the District level machinery, the Indian Council of Agricultural Research/National Horticulture Board (NHB) can make special efforts to achieve this goal.

Many challenges need to be addressed which demand strong support at all levels. Partnerships can be forged with National Agricultural Research System (NARS), the village level workers and NGOs working with farmers. Funding may also be sought from State horticulture departments.

At yet another level, horticultural nursery co-operatives may launch a tissue culture laboratory each, where, the micro plants generated could be fed back into the nursery chain for onward sale. Thus, the system would be a self-sustaining enterprise.

Other sources that can be tapped are

Department of Women and Child Development (HRD)which offers (a)Special thrust on empowerment and training for women (b) Support to training and employment projects (STEP).

Women's Empowerment and National Development

So far, none or very little effort has been made on entrepreneurship development among women farmers. Such development requires concerted efforts. It is useful in empowering women and enabling them to break the barriers that keep them from taking advantages of commercial agriculture, especially, horticulture. There is a tremendous opportunity in horticulture trade at entrepreneurial level for women.

There is a need to focus attention on women, so that benefits of the schemes reach them. If possible, it may be made mandatory that at least 30 per cent of the beneficiary farmers be women.

Horticulture is becoming an increasingly important enterprise in many agro-based economies with increasing demand for fruits. A major revolution in the horticulture industry has been the commercial application of tissue culture biotechnology. Women constitute almost 50% of the total workforce in any country. Traditionally, breadwinning has been a male bastion with the role of women restricted to home keeping, their contribution to society thereof being underplayed. Women could earn economic independence if they launch agri-cottage industries from their homesteads working on their projects during free time. The concept outlined above would give women meaningful employment to help build national food security, thereby, imparting to them a modicum of self-esteem and improving their standard of living. By doing so, they would not only enhance their self-worth but also contribute to national development.

Reference

Arve, A.D.2000. New Rural Enterprise based on Plant Tissue Culture. In: *Taking Biotechnology to Indian Farms, Proc. Natl.Conf.* July 26-27, 2000, Ahmedabad. 56pp.

7

Role of Women in Popularisation and Adoption of Bio-control Technology

CHANDISH R. BALLAL

ABSTRACT

Bio-intensive pest management modules have been developed by the Project Directorate of Biological Control (PDBC) for management of pests on cotton, sugarcane, rice, citrus and several other crops. These modules lay emphasis on release of bio-control agents like parasitoids, predators and pathogens and reducing chemical pesticide applications to the possible extent. The pre-requisite of any bio-control programme is to have a large-scale supply of beneficial agents. It is at this juncture that women can play a major role. If the biocontrol strategy has to become popular, there should be reliable sources, which can assure timely and adequate supply of quality bio-agents for large-scale field releases. It would be most appropriate to educate farm-women on the importance of natural enemies and to provide them with incentives to start bioagent production units. Again, if the concept of biological control with the involvement of women has to succeed, massive educational programmes should be initiated. Women can become active participants at the village level mass production units, which could cater to the needs of farmers of different regions of our country. This could solve the problems encountered in long distance shipment of bio-agents. Besides creating awareness among women, these insectaries/production units can provide job opportunities for the rural and urban women. The stake-holders, who can take up the responsibility of spreading the message regarding the participation of women in biological control, are the policy makers, researchers, extension workers and the Department of Women and Child Development.

INTRODUCTION

The advancement in agricultural technology has brought about remarkable changes in the agricultural sector. These changes have been accompanied by excessive use of pesticides. World-wide, there are 500 species of resistant insects, mites and ticks compared with only 25 in 1955. Coupled with this, is the well-publicised environmental effects, such as toxic residues on produce, destruction of beneficial insects and other non-target organisms, and human poisoning. The World Health Organisation (WHO) estimates that world wide over a million people are poisoned with pesticides each year and up to 2 percent of cases may prove fatal. At this juncture, bio-pesticides offer an alternative method of control that do not seem to provide the rapid development of resistance in the field, leave little or no toxic residues and are generally harmless to beneficial insects and other non-target organisms.

Bio pesticides in Agriculture: We can define bio pesticides as living organisms, which can intervene in the life cycle of insect and other pests or plant diseases in such a way that the crop damage is minimized. The agents employed as bio pesticides are parasites, predators, fungi, bacteria and viruses, which are natural enemies of pests and/or diseases. These bio agents can be conserved, preserved and multiplied under laboratory conditions or in commercial production units for field release against target pests or diseases. A major benefit in the use of bio-pesticides is that they are safe for use by human beings and there are no reports regarding hazards caused due to use of bio-pesticides, while there are innumerable instances of poisoning due to chemical pesticides on non-target organisms.

Developing countries would be highly benefited through development, exploitation and use of biopesticides. The production, sale and use of biopesticides in a developing country can provide local employment opportunities, reduce health risks and costs due to chemical poisoning, minimize environmental damage and improve export earnings through reducing chemical residue levels on export commodities. In addition to this there are the benefits obtained through the extra control achieved by preserving natural enemies in crop systems and by maintaining indigenous bio-diversity.

Bio-intensive pest management modules have been developed by the Project Directorate of Biological Control (PDBC) for management of pests on cotton, sugarcane, rice, citrus and several other crops (Singh, 1996a, b; Rabindra & Ballal, 2002). These modules lay emphasis on release of bio-control agents like parasitoids, predators and pathogens reduce chemical pesticide applications to the possible extent. The pre-requisite of any bio-control programme is to have a large-scale supply of beneficial agents. Today, in our country, there is a great demand for bio control agents, but the major problem is with respect to availability of good quality bio-agents at the required place and time.

POTENTIAL BIO-PESTICIDES

In India, several biopesticides have been identified, evaluated and recommended for field release. Biological control systems for agricultural pests on several crops are readily available in the country (Singh, 2001). Technologies are available for the production and use of bio-pesticides like parasitoids (e.g. *Trichogramma*), predators (e.g. Chrysopids and Coccinellids) and microbial bio-pesticides (e.g. Nucleopolyhedrosis viruses, *Bacillus thuringiensis,* entomofungal pathogens, fungal and bacterial plant disease antagonist, etc.) (Singh, 2001; Jalal *et al.,* 2003; Rabindra *et al.,* 2006).

There are several success stories in the field of bio-suppression of crop pests in our country (Singh, 1996b). Biological control has gained maximum acceptance among sugarcane farmers of India. Use of *Trichogramma* species for the management of sugarcane borers is being followed in a major way. Sugar mills have their own co-operative parasitoid production units and have contributed in a big way in adoption of bio-control. In Punjab, a Bio-intensive Integrated Pest Management schedule has been developed for management of cotton pests. By adopting this schedule, the number of insecticide sprays in cotton ecosystem was reduced to five in the entire crop period. Besides, the egg parasitism of bollworms reached up to 70% and development of resistance to pesticides in the case of white fly was not recorded. In rice ecosystem conservation and inundative release of egg parasitoids of *Trichogramma japonicum* and *Trichogramma chilonis* was successful in suppression of borers and leaf folder.

A large number of highly effective disease antagonist isolates have been identified and found effective in the laboratory, greenhouse and experimental field conditions, but rarely in farmer's field. Few commercial products are available. A talc based formulation of *Trichoderma viride* for dry seed treatment of groundnut, gingily, sunflower, cotton, chickpea, black gram, green gram and pigeon pea to control *Macrophomina phaseolina* has been developed and has been distributed to farmers to cover about 20,000 ha annually.

Potential isolates of *Psedomonas fluorescens* have been identified. Their efficacy alone or in combination with other agents like *Trichoderma* spp. has been studied against wilt diseases of pigeon pea, sunflower, etc. Formulations of *P. florescens* with various carrier materials have been developed, amongst which the talc-based one could give best results.

ROLE OF WOMEN IN POPULARIZATION AND UPTAKE OF BIOLOGICAL CONTROL

Surveys have revealed that rural women in India are aware of the utility of chemical pesticides for management of house-hold and crop pests. But the need of the hour is to measure their awareness regarding the harmful effects of pesticides directly on human beings and indirectly through environment. There is a need to educate them regarding the importance of conserving beneficial organisms through reduced use of pesticides and create awareness regarding pest management utilizing bio-control agents. Farm women are known to be highly receptive, have greater insights, are more critical and can thus play a significant role in envisioning the changes required to build a sustainable agricultural system.

Although the Indian women folk have been playing significant role in agriculture since times immemorial, they have often been relegated to the background. Even today, they are the major partners in almost all the agricultural activities and operations, be it sowing, transplanting, weeding, rouging, picking, plucking, harvesting or threshing. But unfortunately, their contribution has not been fully recorded, recognized and rewarded. In so-far-as plant protection is concerned, women are playing a vital role in weeding, rouging (of pest infested plants/parts) and also in the pesticidal spray operations in every part of the country.

There is clear indication that farmers in India have started accepting the Bio-intensive pest management (BIPM) modules in place of a complete chemical control strategy. Yet, the bio control method of pest and disease management has not become popular due to some constraints in adoption. The most important constraint is the lack of reliable sources, which can assure timely and adequate supply of quality bio-agents for large-scale field use. At this juncture, it would be most appropriate to educate farm-women on the importance of natural enemies/bio-agents and to provide them with incentives to start bio agent production units. Such units could take up the production of those agents, which are in great demand and for which, simple production technologies are available. The agents produced by such units could be utilized by farmers for field use or research organizations for their research activities. The protocols involved in the production of bio-agents are to be followed systematically and on a regular basis. Based on the skills which have generally been identified in farm women, it is evident that they can play a major and effective role in production of bio-agents, provided they are given adequate training and facilities.

Gender stereotypes, which persist in urban and agricultural settings, devalue women's work and, ironically, lead to women doing a disproportionately greater amount of work, which is generally not quantified. It is important to get women involved in productive and quantifiable activities. Production of bio-agents at the local/ village level could definitely be one such activity.

TRAINING FARM WOMEN IN IPM AND BIOLOGICAL CONTROL

A large number of women are being trained in IPM approach under the farmer field schools organised by Central and State Governments. In the FAO-EU IPM Cotton Programme, almost thousand women were trained in farmer field schools in Karnataka, Maharashtra and Andhra Pradesh. Women are also being increasingly involved in the production of bio-control agents in various parts of the country.

An attempt towards creating awareness among farm-women was made by the Project Directorate of Biological Control (PDBC) through a training programme entitled "Village level mass production of egg parasitoids of *Trichogramma* spp. and antagonistic fungus *Trichoderma* spp." which was sponsored by the World Bank under its project on "Empowerment of Farm Women". Twenty five farm women were trained on the methodology of production of two bio-agents. They were given practical training at PDBC and at the village level on in-situ production using simple methods and easily available materials. This village level training programme at village Maliappanahalli created a lot of interest among the farm-women. Some of them were interested in taking up the production of these bio agents for their own use as well for sale in the local market. However, they expressed the need for Government support for the entrepreneurship and creating market for their products.

WOMEN ENTREPRENEURSHIP IN PRODUCTION OF BIOCONTROL AGENTS

At Tuticorn, fifteen farm women belonging to a Tamil Nadu Women in Agri Business and Extension group (TANWABE) at Ankamangalam established a bio control agent production centre in their area (Vimal Kumar, 2007). Subsequently, their life has undergone a drastic transformation. The women have become entrepreneurs supplying bio control agents like *Pseudomonas fluroscens* and *Trichoderma viride* on commercial basis. The unit was set up by the group on an outlay of Rs.5 lakhs extended to them by the Department of Agriculture as revolving fund assistance. It has been equipped with state-of-the-art gadgets such as rotary shaker, distillation unit, hot air oven, vertical autoclave and incubator. The group plans to produce five tonnes of bio agents a year. The products of the group are available in 200 gram sachets. While a packet of *P. fluorescens* is priced at Rs.15/-, *T. viride* is priced at Rs.20/-.

A project on agro-ecological conservation was implemented at Krishi Vigyan Kendra, Pravara Institute of Research and Education in Natural and Social Sciences (PIRENS), Babhaleshwar in Ahmadnagar District, Maharashtra. Bio agents for control of pests and diseases were produced and supplied to the beneficiaries for demonstration. Scheduled caste/Scheduled tribe families were selected to undertake training, demonstration and trials of different bio control agents of pests and diseases (HaNPV, *Nomuraea* spp., *Trichoderma* spp., *Paecilomyces* spp. and *Verticillium* spp.) on various crops. The beneficiaries earned an average income of Rs. 2800/- per acre.

The Department of Biotechnology has initiated several activities under its Women Biotechnology programme. The main thrust of activities include development, promotion and use of biotechnological processes and tools for entrepreneurship development and providing employment opportunities to women, to set up integrated facilities for economic empowerment of urban as well as rural women and to support R&D projects specifically addressing the problems of women or for developing technology packages relevant to them. The DBT could play a very important role by encouraging researchers to take up projects on village level production of bio-agents with a large participation of rural/urban women. The project should not only aim at empowering these women with the know-how on production of bio-agents but also should find an assured market for their produce.

If the concept of biological control with the involvement of women has to succeed, massive educational programmes should be initiated to train women through training aids like pamphlets and booklets (in local languages) with illustrations on useful and harmful insects, protocols to be followed to produce bio-agents and use them in the field, the advantages in using bio agents and dis-advantages of using chemical pesticides. Radio and television talks highlighting these aspects would also be useful in making the concepts clear. Women can become active participants of the village level mass production units, which could cater to the needs of farmers of different regions of our country. This could solve the problems encountered in non-availability of quality bio-agents locally and long distance shipment of bio-agents. Besides creating awareness among women, these insectaries/production units can provide job opportunities for the rural and urban women. The stakeholders, who can take up the responsibility of spreading the message regarding the participation of women in biological control, are the policy makers, researchers, extension workers and the Department of Women and Child Development.

References

Jalali, S. K., R. J. Rabindra, N. S. Rao & C. B. Dasan. 2003. Mass Production of Trichogrammatids and Chrysopids. Technical Bulletin No.33. Project Directorate of Biological Control, Bangalore. 16 pp.

Rabindra, R. J. & C. R. Ballal. 2002. Mass Production of Bio-agents and their use in Integrated Pest Management–Problems and Prospects In: *Resource Management in Plant Protection*. Volume I (Eds.): B. Sarath Babu, K. S. Varaprasad, K. Anitha, R. D. V. J. Prasada Rao, S. K. Chakrabarty & P. S. Chandurkar. pp. 255– 273.

Rabindra, R. J., C R. Ballal & S. K. Jalali. 2006. Utility of parasitoids in Biological Control. In: *Plant Protection in New Millennium* Volume II (Eds.): A. V. Gadewar & B. P. Singh. Satish Serial Publishing House, Delhi. pp. 197-226.

Singh, S. P. 1996a. Biological control in India. In: *IPM System in Agriculture Volume 2 - Biocontrol in Emerging Biotechnology*. (Eds.) Rajiv K. Upadhyay, K.G. Mukherjee & R.L. Rajak. Aditya Books Pvt Ltd., New Delhi, India, pp.216-328.

Singh, S. P. 1996b. Biological control. In: *50 years of Crop Science Research in India - Golden Jubilee Year of India's Independence 1947-1996*. (Eds.) R.S.Paroda & K.L.Chadha. Indian Council of Agricultural Research, New Delhi. pp. 89-116.

Singh, S. P. 2001. Augmentative biocontrol in India. In : *Augmentative Biocontrol– Proceedings of the ICAR-CABI Workshop*. June 29th to July 1st, 2001 (Eds.) Singh, S. P., S.T. Murphy & C.R. Ballal. Project Directorate of Biological Control, Bangalore. pp. 1-20

Vimal Kumar, R. 2007. Biocontrol agent production centre is a boon to them. *The Hindu*. August 26th, 2007.

http://www.dbtindia.nic.in/programmes/biotechsocietal.html

http://agricoop.nic.in/womenagri02.htm

8

Women in Agriculture, Agricultural Research and Extension in India

SHAKUNTHALA SRIDHARA

ABSTRACT

Scarcity of resources, economic deprivation, gender bias in wages, prevention from decision making and using income generated by farm activities has made life miserable and unjustifiably hard for rural women. The plight of rural women can be improved by dissemination of latest technologies in agriculture, creating small entrepreneurial outlets for agri-based products, networking of support services for women, development of new and adoption of drudgery reducing technologies, improvement in rural infrastructure to improve the life style and to promote entrepreneurial opportunities, better access to resources, organizing women's co-operative movement, larger role in decision making, mandatory involvement of women in all agencies and organizations involved in rural development, feedback and follow up of women's programmes and recognizing the success of rural women by government and its highlighting by media.

Women scientists in agriculture are far less than in other branches of science. In spite of the demands of research and extension activities in agriculture being physically strenuous with lot of field work and village visits, women can contribute effectively even in these fields. Women by an evolutionary perspective are nurturers. Taking care of the unborn, babies, young children and elders is their inborn asset and strength which makes them ideal to raise new varieties of crops and breed better strains of crops and races of livestock and pet animals. Even the inexhaustible patience required by genetic engineering to evolve modified crops may be a better field of research to woman than other branches of sciences. As in other professions, women scientists and extension workers in agriculture face several problems because of their dual role as home makers-mothers and professionals. Apart from detailing the problems of farm women and women agricultural scientists, recommendations to alleviate their problems are summarized in this review. Steps to make extension programmes more focused, targeted and gender sensitive are also highlighted.

INTRODUCTION

Research and education in agriculture are vital for the prosperity of the country as we are an agro based economy, society and culture. Agriculture and its allied activities form the core of Indian economy as they not only contribute significantly to national GDP but also involve the largest proportion of country's population through direct or indirect employment particularly rural employment. The role of women in agricultural production is two fold: on one hand women scientists are developing technologies for increased agricultural productivity and transferring them to farmers including farm women. At the receiving end farm women become students, learning the improved technologies as well as adapting them in their farming activities. The potential, challenges and problems are similar for both categories of women to certain degree when we take into consideration the responsibility of family management. But they differ drastically in the duties they discharge as professionals. The two categories of professional responsibilities and constraints are dealt separately in this paper.

I. Rural Women In Agriculture

The Plight of Rural Women (Table 1): Women of India, numbering 495.7 million constitute 48.27% of total population of which 72.72% live in rural areas. Their literacy level is only 46.7%. Female labour force in rural areas is 31.56% with most of them (87%) being employed as agricultural labourers and cultivators (Suleiman *et al.,* 2003). Farm women of all ages shoulder the major burden of subsistence agriculture. They carry out 75-80% of farm work in family farms while their contribution in hired agricultural labour is 50-66%. Some agricultural activities such as ploughing, sowing, transport and sale of agricultural products are exclusively male responsibilities where as raising nurseries, transplanting, manual weeding, harvesting in small farms, threshing and storage are exclusively women's activities. Other activities are shared by women and men. In addition women are adept at plant breeding and managing natural resources such as soil and water. They are experts in domestication of wild species and play a significant role in selecting and storing seeds for future crops. Being cooks of the family they not only have detailed knowledge of specific traits of crops but also exhibit strong preferences. They are well aware of the texture and taste of a grain, its shell life, cooking property, nutritive value, alternate uses etc.

Table 1: Pattern of women employment in agriculture and animal husbandry in India (Anon 2001, from internet)

Type of activity	Percent/millions employed
Total employed	75-80% (Men, 63%)
Agricultural labourer	50-66%
Farming (self employed)	48%
Dairying	75 million (Men, 15 million)
Animal husbandry	2 million (Men, 1.5 million)
House hold chores	95%

In case of livestock also more than 90% of the work related to animal care is done by women. In diary 75 million women are working forming 75% of work force compared to 15 million men. Similarly in animal husbandry two million women are engaged compared to 1.5 million men. Equally daunting is household chores ninety five per cent of which is discharged by women. This dominant role of women in agricultural production and house keeping is not limited to India alone but is seen in other countries too including USA (Mies, 1987; Legg, 2002).

Scarcity of resources, economic deprivation, gender bias in wages, decision making and inability to use income generated by farm activities has made life miserable and unjustifiably hard for rural women. Barring the women of rich and affluent farmers, rural women work either on their own farms or become agricultural labourers to raise supplementary income where caste and family restrictions do not exist for such outside employment. Those who cannot go out to fields, farms and forests engage in income generating activities at home. Generally rural women participate in agricultural operations like manuring, weeding, land preparation, hoeing, sowing, applying manure and fertilizers, harvesting, threshing, storage, transporting produce from farm to home and animal care. In case of animal husbandry, they bring fodder from the filed, cut the chaff, feed and water the livestock, clean and bathe them; clean the cattle shed, milk the cows, prepare milk products, and look after the calves, pregnant and mulch animals. Women from poor households engage more in field activities while those from higher incomes devote more time to household activities. The plight of landless labourers is the most pathetic with heavy burden of work in cultivation including land preparation apart from domestic chores.

As women irrespective of land status of the family have to slog for 14-18 hrs in productive and physical labour of different kinds on and off the farm, the energy demand for such physical strain is enormous. This renders women from below and subsistence level susceptible to malnutrition and disease leading to lower expectation of life. Although the plight and drudgery of rural women is existent from time immemorial, scientific data and remedial measures on these aspects are not adequately gathered and addressed. The few studies carried out highlight the following issues:

When assessed on a four-point scale i.e. easy, moderately difficult, difficult and very difficult, the farm activities carried out by women in Meghalaya namely transplanting, weeding and harvesting, majority of women found these activities as difficult or moderately difficult (Singh *et al.*, 2001). In the neighbouring Assam the Karbi tribal women, hailing from poor income and marginal farming families were involved in the selection of seeds, hulling and cultivation only and perceived these activities as light (Saika & Das, 1998). A detailed study on farm women in hilly districts of Uttar Pradesh revealed that they carried out 16 hours of household and farm work with no time for personal care. They reportedly spent 8-10 hours in farming work such as sowing vegetables and cereals, harvesting, cleaning, husking and storage (Ahuja *et al.*, 1999). The physical cost of farm work for women in Uttar Pradesh as measured by Average heart beat (AHB), Physiological cost of work (PLW) and Energy expenditure (EE) was found to be correlated with load carrying and winnowing activities (Chauhan & Dayal, 2004).

Drudgery reduction: Inventions such as paddy threshers, winnowers, sprayers, harvesting tools, parboiling units, maize shellers, dal making machines (Seth & Bilgi, 2002) are available to reduce the drudgery of women but much more remains to be done to make life less sternuous for rural women. In many parts of the country women even now spend considerable time and physical effort to fetch potable water, fodder and fuel wood. Their plight can be made less laborious by digging wells in the centre of the village, improving manual transport aids such as handcarts, improving cooking stoves, increasing biomass production to meet fuel needs and planting of fast growing fodder in common lands.

Girl Child Labour: A major concern throughout the developing world is the issue of child labour in contract farming. With about 80% of working children employed in agricultural sector, India has the dubious distinction of one of the major users of child labour in Asia. The majority of these child labourers are girls. They are preferred to boys for their docility, obedience and nimble fingers. There is an urgent need to take a gender perspective to address the whole question of changing agrarian production structure under contract farming. Further the working conditions for child labourers should be improved. So also their education and skills so as to improve their economic condition and prevent unjust exploitation (Singh, 2003).

Female children as unpaid contributors have considerable role in household production. They contribute substantially as helpers in subsistence farming, live stock management, gathering fuel wood, collecting water and acting as surrogate mother when the mother is busy in farm and other livelihood activities. By virtue of these various activities they qualify as essential components of rural labour force. But their contribution is ignored by standard estimates of labour force, which takes into account only the 15–55 age groups.

Gender Differences and Gender Bias: Gender differences exist in both work carried out by rural women as well as in wages they are paid. Farm women in Andhra Pradesh were categorized into three types namely farm, landless marginal and small farm women. Farm women utilized time for agricultural activities, land less for home management and the marginal farm women for dairying and miscellaneous activities (Sashidhar *et al.,* 2001). In Punjab, the level of women's participation in agricultural activities was low. Rural women engaged mostly in household activities. Similar pattern of higher attention to household work was seen in Uttar Pradesh but considerable amount of time was also spent in dairying and agricultural activities. Dairying activities included fodder chaffing, washing cattle and dung cake preparation while the agricultural activities undertaken were fodder transportation from fields to homes, harvesting and irrigation (Kumar & Fulzele, 1998). In Meghalaya women mainly carried out transplanting, weeding and harvesting while men did the ploughing and threshing (Singh *et al.,* 2001). Similarly in West Bengal rural women more actively indulged in household activities than in poultry keeping and other activities related to agriculture. The conservative family structures and low education levels of women were the reasons attributed to this disparity (Kar *et al.,* 2002). In Kerala majority of farm women "solely" participated in sowing/planting and harvesting of vegetables, half of them supervised plant protection operations, half of them jointly decided the type of weed control to be adopted, manure/fertilizer to be applied and irrigation management (Suresh Kumar *et al.,* 1998a). In livestock management, majority of farm women solely took responsibility of maintaining cattle sheds, feeding poultry birds and hatching of eggs/chicks. But when it came to the care of sick animals, type of milk product to be prepared, hatching of eggs, marketing of eggs and milk, the decisions were taken jointly by farm women and men (Suresh Kumar *et al.,* 1998b). In Karnataka majority of female farm labourers participated in weeding, harvesting, winnowing and transplantation in rain fed areas. Digging of trenches was rare. In the irrigated area, grass cutting was the major job for women labourers (Geethalakshmi *et al.,* 2002). In contrast, the tribal women (santals, mundas and oraons) of West Bengal participated more in subsistence activities such as trading firewood and other forest produce and less in other activities such as agriculture, animal husbandry, agricultural labour, liquor trading, hunting, fishing, capture/collection of various animals and plants which was carried out by men. However, women were completely responsible for rearing of animals (Pal & Seal, 2005).

Mechanization of Agriculture and their Negative Impact on Rural Women: With irregular rains, droughts, decreasing land availability and other resources, attractive salaries in cities, coupled with increased debts in villages a large number of male farmers are migrating to cities and in some states like Punjab and Haryana to cities outside India. This has resulted in women becoming increasingly involved in agricultural activities in what is termed as "Feminization of agriculture". This is discussed in detail in a later section. In this changed situation more and more poor women are employed as agricultural labourers. Although this changed labour arrangement has lead to slight increases in the income of women labourers, they are paid as usual less than men. Their working conditions are poorer and their bargaining power is limited.

Modernization of agriculture has not favoured farm women in most developing countries including India. Mechanization of agriculture by installation of pump sets, tube wells, diesel engines and improved technology by the introduction of high yielding varieties, fertilizers, and pesticides are the major contributors to modernization of agriculture. Mechanization includes use of hullers, combine harvesters, tractors, rice transplanter etc. In general these improvements have lead to increased agricultural productivity and decreased drudgery but have adversely affected the women from lower castes, lower income groups, and landless families or with small land holdings by reducing demand for labour. Mechanization has occurred for activities usually carried out by men while women continue to toil in labour intensive chores, which also happens to be most monotonous and lowly paid. This is seen in activities such as weeding, harvesting, load carrying, food processing etc. In addition, whenever employment opportunities decrease in agriculture, the most affected are women.

Introduction of weedicides and pesticides has deprived a large number of women labourers of their meagre source of income. Gender targeted studies in Maharashtra indicated that technology increased workload of women farmers and expanded the employment prospects of female wage labourers. But technology also enhanced the unequal power relationship between men and women. There was a tendency for decreased resource allocation to women but they gained marginally in benefit distribution. Simultaneously women were pushed into domestic sphere and men dominated marketing (Kolli & Bantilan, 1998). A parallel study in Tamil Nadu during 1977-1999 revealed significant increase in the wage rates of women and widening of gender gap in wages. The earlier stages of green revolution saw a great increase in female labour absorption in agriculture but later phases were characterized by deceleration of female labour absorption in agriculture due to introduction of weedicides (Ramachandran *et al.*, 2001). In contrast, drastic increase in demand for female labour seen in Andhra Pradesh consequent to green revolution during 1971-72 and 1997-98, continued while mechanization and other technological improvements in agriculture reduced the demand for male labourers severely (Subramanyam, 1999).

Farm Women and Pesticides: With approximately 2.5 million tons of pesticides applied annually worldwide, both male and female farm workers are exposed to these dangerous chemicals. The exposure can be direct when they are functioning as pesticide applicators or indirect during harvesting, planting and soil preparation. Generally the risk of mortality as a result of this exposure is twice that of risks in other employment categories. With 43% of total workforce in agriculture comprising of women, the threat of pesticide toxicity is alarming. Some pesticides affect not only women agricultural workers but also the intra-uterine growth of foetus. Although there is sufficient data to link pesticides with leukaemia, lymphoma, a plastic anaemia, soft tissue sarcoma and cancers of the breast, brain, prostate, testis and ovaries, women are more susceptible as a number of reproductive impairments have resulted due to pesticides in women. Herbicides such as 2,4-D and 2,4,5,-T are known to disrupt estrogen cycle in women, Aldrin, dieldrin, chlordane and toxaphene are known to disrupt reproduction cycles. There is also evidence of increased risk of birth defects from maternal exposure to pesticides. Worldwide reports indicate presence of DDT and its metabolites in human breast milk including the fact that in Delhi, a breastfed infant receives 12 times the acceptable limit of DDT (WEDO, 1999).

Although there does not seem to be many studies on pesticide hazards to farm women in India, the adverse effects would be similar to other reports, may be the degree varies. The relevant solutions includes ban on certain dangerous pesticides, enacting of strict regulations on the sale, labelling of all pesticides, educating the farm women on hazards of pesticide use, alternatives to pesticides, adoption of integrated pest management, reduction in pesticide use etc.

Decision making: Women's role in decision making is marginal. All the major decisions within the family are taken by the men of the household. Although women are allowed to be independent in child rearing and house keeping, she is not consulted in matters dealing with investment, expenditure and development activities. Very few studies exist about the extent of decision-making by farm women in India. In Uttar Pradesh, husbands were the key decision makers in presowing activities and joint decisions by husband and wife were made in sowing, post sowing, harvesting and post harvesting activities (Kunwar, 2004). In Andhra Pradesh very few decisions on agriculture and horticulture are taken by women (Reddy & Rashid, 1977). The situation is better in the case of livestock. In almost all the studies most crucial and economic decisions like selling of produce, purchase of inputs and animals were exclusively decided by men. The scenario about spending farm income is more pathetic. In Uttar Pradesh only 7% women spent money earned, made 1.22% decisions relating to taking credit for farm activities and 2.44% made decisions regarding taking loans for live stock management (Roy & Goyal, 1998).

Feminization of Agriculture: Added to the already existing workload, of late more and more men are withdrawing from agriculture due to varied reasons in different countries leading to feminization of agriculture as already pointed out. The reasons could be economic, and landlessness globally, and epidemics in African countries. The ultimate result is more burden on women and their increased participation in agriculture. Thus woman is not a secondary worker in agricultural development but has become the mainstay. She has to be at the centre stage of policy decisions. Her status has to be strengthened so as to enable her to play her role at least as equal partner if not dominant. This is possible by educating rural women, by closing the gap between research institutes and women farmers, by training more women agricultural experts, securing property rights and access to finance and improving nutrition. In addition to imparting skill, information technology and motivating towards better health and self-cure, her enormous workload has to be lightened.

What Is Needed to Improve the Life of Rural Women?

Compared to the magnitude of women's involvement in agriculture, the studies on the issue are scanty, were conducted under lot of limitations and do not adequately reflect the role of farm women in various agricultural activities so as to understand their plight completely. Detailed data on the time allocation to different domestic and farm activities, their potential, constraints faced by them are essential for economic analysis and policy decision in agricultural planning, rural development and reform in agriculture. This needs to be done across the country so as to evolve area specific agrarian planning and reform as socio-economic conditions are vastly different from region to region.

Empowerment of women in agriculture as in any other sphere of development is demanding and challenging , and involves overcoming barriers so as to enable her to improve and shape her life comfortably in a sustainable fashion. Unfortunately the contribution of women in agricultural production has not been given its due status due to lack of interest, information and equally by neglect. Consequently women are denied their rightful status as active producers in agriculture and are thus unable to access resources and services as a result of which they are getting marginalized. To increase productivity at a faster rate, to reduce environmental degradation and to conserve resources as well as to enhance their standard of life, women in agricultural sector have to be empowered completely and very fast. This can be achieved by enhancing their awareness, knowledge, skills and efficiency to use technology. This calls for, first and foremost due recognition of the role and contribution of farm women followed by sensitization of extension services towards their specific problems, capacity and capability leading to their total integration in agriculture.

Progress Achieved

The government of India has initiated several programmes to empower rural women. A number of gender neutral and simple devices/equipment/machines for crop production and processing have been developed. The adoption of these devices will not only enhance agricultural production but also free them from drudgery to a great extent allowing them to utilize their time in more fruitful ways. Simultaneously they are encouraged to take up income generating agricultural based activities such as dairying, animal husbandry, poultry, goat keeping, rabbit rearing, beekeeping, floriculture, horticulture, fruit preservation, post-harvest technology, value added food products etc. Additionally location specific, income-generating ventures are taught and encouraged such as mushroom cultivation, low cost nutritional foods, coconut processing, coir sector, lac cultivation, agro processing etc.

To lessen the burden of household work, kitchen appliances such as improved chulha, biogas, solar stoves, solar heaters, photovoltaic energy operated appliances are fabricated and made available at low cost and on instalment basis.

To create employment opportunities including self-employment and income generation, various economic incentives and investment opportunities are available for rural women. These include special financial schemes for women in various financial institutions, National Bank of Agriculture and Rural Development (NABARD), Krishi Vignan Kendras (KVK), Co-operative societies and regional rural banks. They provide loan opportunities for women without any security up to few lakhs at low interest rates.

Research in the institutes of Indian Council of Agricultural Research (ICAR) and State Agricultural Universities (SAUs) are focused on relieving women of the drudgery of agricultural activities by providing time and labour saving devices. They are imparted vocational trainings enabling them to undertake different vocations. There is rethinking in the central policy. In the forefront of such research is the National Centre for Research on Women in Agriculture, Bhubaneshwar (NCRWA) and Krishi Vignana Kendras. The NCRWA is concentrating on women specific technologies under different farming systems. The major projects in the Institute are:

- Standardization of women specific field practices
- Occupational health hazards
- Reducing drudgery of women in agricultural operations
- Improvement of farming system suited to farm women
- Ecofriendly pest management technologies for women to cultivate vegetables.
- Technological needs in empowering women in rural agriculture (Self help groups)
- Improvement in storage practices

Economic empowerment through Self-help groups (SHGs): Vast numbers of SHGs have been established across rural India through NGOs, government agencies and rural banks. These function as self-reliant autonomous local financial intermediaries. Eighty five percent of the members of SHGs are women who have proved themselves to be better savers, borrowers and investors. Most of them are from lower castes and other disadvantaged families. Basically the SHGs mobilize funds by investing savings of members which is then given as loan at low rates to members. The interest earned is ploughed back to the SHGs to widen the group and its activities. An evaluation of SHGs in terms of benefits towards economic and social empowerment in all the states of the country spread across north, south, east and western region has shown that the system has successfully achieved its goals. It was found that about 59% of the sampled households had registered an increase in assets, average income increased by 33% and 47% of borrowers moved above the poverty line. Social empowerment was reflected in improved confidence levels of members, the way they were treated by the rest of family members, enhanced communication skills and other behavioural aspects. SHGs in southern and western region fared better compared to the other two regions (Puhazhendhi & Satyasai, 2002). The organization and functioning of the group with emphasis on social cohesion and election/selection of the group leader in rotation created a sense of responsibility in members. Stress on both consumption and production while lending paved the way for productivity and income generating activities. (Numboodiri & Shiyani, 2001). A higher degree of success of SHGs was seen in Tamil Nadu where the lending pattern shifted from a predominantly consumption to production purpose. This in turn resulted in income generation, savings and empowerment of women. Intermediation of NGOs and SHGs reduced the transaction cost up to 59% and the recovery rate of 92% was extremely encouraging (Puhazhendhi, 2000).

Namboodiri & Shiyami (2001) consider the limited scope for future growth in membership as the main weakness of SHGs. Another negative observation was the emphasis on consumption loans, which limited the scope for income generating activities. The most important drawback was that the SHGs do not have a legal status.

Further Actions Suggested

Apart from attempts to educate both formally and informally (transfer of technology), property rights, access to credit and better health through balanced nutrition, there are a few important measures that will improve the life of farm women. More emphasis should be placed on reducing the manual work of women in villages and adoption of drudgery reducing technologies and machines. The American example of 100% rural electrification relieving farm women of household drudgery is an example worthy of emulation. Though it looks far fetched in the present day Indian context, women freed from carrying potable water, manual washing, and long hours in the kitchen, walking long distances to fields etc, can be much more productive, healthier and happier through rural electrification contributing to a stable society. It is not an impossible dream. Even as recent as 1930, only 10% of rural America was electrified. With 100% electrification through a co-operative movement called Rural Electrification Association, within decades American women were empowered. There are other issues too apart from electrification - good roads, readily available, safe, cheap fuel to cook, potable drinking water etc.

The United Nations places access to information as the third most important issue facing women globally after poverty and violence against women. Apart from basic education to girls and women in villages, there is an urgent need for percolation of information into women's education both formal and informal. This can be easily done by giving emphasis on women's responsibilities in agriculture, equal footage in programmes meant for farmers on TV, radio and other channels of communication.

Most importantly there is a great need to form co-operatives to market not only the already saleable products but also to raise a number of new products. For example there are 130 co-operatives in USA that are found and run by women. Every thing from sweet potato to crab meat products, quilts, weavers etc are managed by co-operatives run by women. AMUL in Gujarat is the finest example of successful women's co-operative movement. It is time the movement spreads to other rural products and activities.

Recommendations Made to Empower Rural Women are:

- Dissemination of latest technologies in agriculture.
- Creation of small-entrepreneurial outlets for agri based products.
- Networking of support services to women.
- Emphasis on adoption of drudgery reducing technologies.
- Improvement in infrastructure like road, electricity, telephones, potable water etc in rural areas for increasing entrepreneurial opportunities.
- Control over resources.
- Need to organize women's societies and women's co-operative movement.
- Access to family financial resources and larger role in decision making.
- Creation of entrepreneurship development centre in agricultural universities.
- Mandatory Involvement of rural women in all agencies and organizations engaged in rural development.

- Feedback and follow up mechanisms of women's programmes.
- Highlighting of rural women's success in media.
- Recognizing outstanding rural women entrepreneurs and farmers at state and national level.

II. Women as Agricultural Professionals

In a landmark study on professional women in agriculture, several interesting facts about the aptitude, success and problems of women agricultural graduates were revealed (Table 2, Rama Rao & Shenoy, 1998) emerged. The number of women graduates in agriculture rose steeply from 387(3.9%) in 1980 to 3773 (21.6%) in 1996 with only 14.9% of them coming from rural backgrounds. Girls from urban schools constituted 85.1% of agricultural students. The reasons for choosing agriculture were: it was challenging, offered a chance to work with nature and helped to tap resources for the economic development of the country. Majority did not find fieldwork strenuous. Only 12% found it problematic mostly due to lack of practical training and guidance, negative attitude of colleagues and distance to the field. Home science was the most preferred specialization followed by horticulture and extension. After graduation employment opportunities were excellent with 72.6% getting jobs. For the remaining, unemployment was a choice as they were averse to frequent travels and intense field work.

The women agricultural graduates mostly preferred to work in agricultural institutions followed by development departments of the government, research institutes and private sector. Financial institutions were the least preferred. Majority of employed women felt that they have to do better than men to get equal professional recognition, felt family is of primary concern compared to career and family support is a key factor to a successful career. If they have to work in field stations where infrastructure is poor, they opined that it is better to work in such places early in career, before marriage because family obligations are minimum at these stages. The major reason for not getting jobs in private sector was gender bias. The number of females at top management was negligible.

Table 2. Highlights of study on professional women in agriculture (Rama Rao & Shonoy, 1998)

S. No	Parameters studied	Findings
1.	Increase in women agricultural graduates	From 387(3.9%) in 1980 to 3773(21.6%) in 1996
2.	School background of agricultural graduates	14.9% from rural background; 85.1% from urban background
3.	Reasons for preferring agriculture	(a) Challenging (b) Working with nature (c) Resource utilization for country's welfare
4.	Difficulties in field work	88% did not find any difficulty; 12% found it difficult
5.	Preference for specialization	Home science (40.9%)> Horticulture (18.9%)> Extension (15.7%)
6.	Preference for employment type	72.6% employed. Educational institutes (39.7%)>Govt. Depts. (19.2%)> Res. Inst. (7.9%)> Private sector (4.2%) >Finan. Inst. (1.1%). Unemployment was due to choice.
7.	Gender bias if any	Majority felt that they had to work harder. Gender bias in selection in private sector
8.	Representation at top level management	Negligible

Rama Rao & Shenoy (1998) made the following recommendations to alleviate existing problems and to make working conditions in agriculture sector attractive for agricultural graduates and women agricultural scientists:

(1) More weightage should be given to students from rural background while admitting to agricultural courses as urban girls are generally hesitant to serve in rural areas (2) Reservation for women in scholarships, admissions, nomination to awards, curricular and co-curricular activities. (3) The atmosphere in places of employment

should be made gender friendly. Women employees are to be provided with facilities like transport, accommodation and spouse employment. (4) To minimize gender bias, one women member should be included in all committees such as selection, recruitment, promotion, grievances etc. (5) Modification in work structure by introducing flexible timings, easy leave option and contract jobs (6) Policies have to be evolved to provide spouse employment and preferential placement of women. (7) At workplace and training institutes, essential infrastructure facilities like exclusive boarding, lodging, washroom etc should be provided (8) Women professionals should be encouraged to participate in conferences and training programs (9) Establishment of women cell in all workplaces should be mandatory (10) Networking of agricultural professional women at organizational and national level should be taken up on a priority basis.

II (A): Women Scientists in Agricultural Research

In India the proportion of girls choosing agriculture is far less than other faculties. Generally percent of women in post graduate and PhD programmes was 42.5% and 37.2% respectively in science subjects. If the survey data is further analyzed medicine had highest percent of women in research (29.3%) followed by Engineering/Technology (16.5%). Agriculture (14.6%) and veterinary (14.5%) were poorly represented by women (Menon, 2005).

In another study, similar pattern of lower women representation was seen in agricultural research. Department of Biotechnology has the highest percent of women scientists (37.8%) followed by Indian Council of Medical Research (ICMR) (27.3%), Council of Scientific and Industrial Research (CSIR) (13.0%), Department of Atomic Energy (DAE) (16.5%) and the central universities (15.8% to 16%). Indian Institute of Science, Bangalore had 6.6% academicians and 9.7% of women scientists. Indian Council of Agricultural Research had a pathetic cumulative representation of 6.6% women scientists of which 10.4% Assistant professors, 6.2% Associate Professors and a mere 3.5% Professors, nearly 50% less than the least percent seen amongst the other institutes (6.6% of scientists in IISC).

Agricultural Extension

Extension essentially involves transfer of knowledge and material towards widespread adaptation of newer technologies to achieve rapid agricultural growth. The knowledge acquired through research should be applied to meet different demands and needs of the farmer. Since the farmer is involved in several activities simultaneously it is essential to executive holistic demonstrations. The ICAR through its institutions and programmes demonstrates the value of various research outputs as technologies helpful to farmers and agriculture. In other words ICAR develops principles and programmes for extension. These demonstrations are essentially through Zonal centres and Krishi Vignana Kendras. The creation and management of these extension centres are under the perview of the Department of Agriculture and co-operation of the state governments. The agricultural extension research has been transferred to SAUs (State Agricultural Universities).

The extension education has evolved through various steps starting from community development, human and community development in fifties with a shift towards technology transfer in sixties characterized by intensive agriculture programs, high yielding varieties programme and farmers training programmes while seventies onwards the approach was by training and visit (T&V) extension management system.

The twenty first century saw a new orientation in agricultural extension with number of new schemes launched. The most important ones are World Bank assisted National Agricultural Technology Programme, Agricultural Technology Management Agencies (ATMA) at district level, State Agricultural Management Training and Extension Training institutes (SAMETI) at gross root level, farmers advisory committees and centres at block level, Agri-clinics and Agri-Business schemes where agricultural graduates are trained to provide extension services to farmers on payment basis and agricultural extension through farmer's organizations and NGOs.

II (B): Women in Agricultural Extension

Extension schemes are still male dominated. Female farmers are reported to receive only 5% of all agricultural extension services worldwide. Compounding the problem is the fact that only 15% of world's extension experts are women. Extension programs more often ignore women component of target audience. Most programs overlook the differential roles and responsibilities rural women hold compared to their husbands and men folk in villages. Thus extension programmes have to be more focused, targeted and gender sensitive. This can be achieved by:

(a) Improving access to extension and training through (i) Extension policy re-orientation, recognizing farm women as extension clientele (ii) Training men and women extension staff on women's role in agricultural and rural development so that it meets the needs of rural women (iii) Training of women in decision making in the context of farm and home management (iv) Training of women in agricultural marketing, post harvest processing, on farm value addition and market requirements/demands.

(b) Redesigning extension services to reach women farmers through (i) Sensitization of extension personnel towards role and contribution of women in the total agricultural system (ii) Increasing proportion of trained female extension workers to ensure at least one third of all extension workers are female. (iii) Improving communication between women researchers, marketing agencies and extension workers. (iv) Developing appropriate extension methodologies that recognize the multidimensional role of women and the socio-cultural barriers in which women farmers operate in the society.

(c) Expanding the scope of women extension workers through (i) Re-examination of all service cadre rules for hidden gender biases (ii) Improvement of female admission to agricultural courses (iii) Increasing incentives such as scholarships and stipends for more women to take up UG and PG courses in agriculture and allied subjects. (iv) Redesigning agricultural curricula to include women's concerns (v) Ensuring that women are adequately represented in all higher training programs inside and out side the country (vi) Taking into account women's time, mobility and cultural situation into the curricula of extension training.

DISCUSSION

While discussing the role of women in agricultural research and extension, in this review emphasis is laid on two factors (1) The gender disparity issue in farm women and women agricultural scientists

(2) Highlighting feminine characters that facilitate research and extension in agriculture.

Gender bias in all fields and at almost all levels of human life is a universal truth. This is partly due to biological reasons and partly due to social attitudes. Women, for that matter females of all species of animals are endowed with biological characters, bodily morphology and physiology to nurture the young inside/outside the body, to take care of them till certain stage and in many cases teach them the lessons of survival. Nature has predetermined this role of propagating species and has bestowed the female accordingly. In case of human beings responsibility of taking care of the young demands a great deal of time and energy of the mother leaving little time to persue intellectual interests at least during certain periods in her life. This is not an issue of choice. It is also controversial. Society in general and men in particular will not easily accept women giving up motherhood in pursuit of intellectual excellence. May be it is a biological necessary too as the female anatomy, physiology, hormonal profile and functioning is evolved towards procreation. Any deviation from this, by way of not giving birth and rearing young may result in emotional imbalance with unpleasant repercussions.

Rural women contribute much towards agricultural activities on farm and in post harvest storage, much more than men. Still they are never conscious of gender discrimination. The idea that these women do not complain against bias is due to the fact that they are independent, not employed by any organization and are free from rules and regulations. This does not seem to be an acceptable explanation. Because women labourers slog much more in the field and get paid less for the strenuous jobs they discharge. They actually balance the triple burden of field work, home and social prejudice. But there are no complaints, no protests. Instead they manage a well adjusted life of domestic and on farm responsibilities. It is a philosophical acceptance that life is a responsibility, home and contributing to family income together.

Women Scientists: Motherhood, family management and societal attitude affect women's pursuit of scientific excellence. Women are traditionally considered and expected to be child bearers, homemakers and men as breadwinners irrespective of their education and profession. It is also an evolutionary truth. But technological and social evolution have freed woman from the drudgery of housekeeping while child bearing and rearing demand women's attention and involvement for only a limited period. Most importantly there is a change in the thinking of both men and women. Women have gradually realized the potential for innovative thinking and contributing to science. Men too, grudgingly have started acknowledging the inherent capabilities of women for research and professional excellence. We seem to be going through a transition in our social evolution as far as women's role is concerned. Many women have adapted beautifully to the dual role of career scientists and housewives. But due to the comparatively less time and concentration devoted to career in this balancing act, it is but natural that women fail to reach the top of the research hierarchy i.e. heading their institution, academies, winning awards, quality and quantity in publication etc. Excellence in science is a full time, dedicated, all consuming passion that very few women can afford to indulge whatever the incentives, encouragement the government or society gives. True, there are gender insensitive practices and sexual discrimination in many organization and institutes. Even when these issues are eliminated or minimized it is impossible for women to realize their full potential because of their dual responsibilities of home making-child rearing and profession. There are instances of women refusing career advancements in preference to domestic responsibilities.

Such being the situation, the prospects for women in agricultural research and extension is much more demanding. With the exception of biotechnological based laboratory research, to some extent plant genetics, almost all areas of agricultural research i.e. agronomy, plant breeding, pathology, entomology, soil science, seed technology, agricultural machinery, post harvest technology etc involve vigorous field work. The ultimate tests of farm trials wherein a successful research concept is vigorously, repeatedly tested in field conditions in villages is a test in physical endurance, patience and tolerance to which very few women can cope up, given their priority to family responsibilities.

On the other hand some of the so called feminine attributes are advantageous in agricultural research and training. Women are nurturers by nature. Taking care of the unborn, young, children and old is a woman's inherent asset and strength. These characters are best suited in plant and animal genetics, and breeding. The patience, the caring attitude and protective nature coupled with her inborn capacity to nurture living things makes her an ideal scientist to raise new varieties of crops and breed better strains and races of livestock and pet animals. Even the inexhaustible patience required by genetic engineering, to evolve modified crops may be a better field of research to woman than other branches of sciences.

Women Extension Workers: The extension education component of agricultural research is equally demanding. Physically it may be less strenuous. Still it involves transferring technology in rural areas involving considerable travel to ill connected, rough terrains. Added to this is gender bias of a different kind. Rural societies are still to accept gender equality. Almost all of them, both men and women basically believe that a women's role is subservient and supplemental to men. They believe that there is nothing wrong in accepting that women should be servile to men, and she should be content with the twin responsibilities of home and work, it is her fate and nothing can be done about it. They unconsciously judge women agricultural scientists and extension specialists as less capable and less knowledgeable about Agricultural technology. Imparting technology through extension education to people of such attitude calls for tremendous grit, determination, patience, understanding and compassion.

Secondly extension education itself is challenging what with farmer endowed with traditional knowledge, rigid mentality, doubtful of urban influence etc. However over the years, the higher yields, expanded irrigation, wasteland development, watershed management, multiple cropping, increased diary income all achieved with the help of personnel of extension departments, and multimedia propaganda have fine tuned farmers to accept and adopt rewer farm technologies. But when women get into the act of extension education, there is initial cynicism and doubts. First and foremost is the inherent suspicion of farmers about urbanites. When an urban educated, agriculture woman scientist tries to teach improvements in farming be it seeds, fertilizers, pesticides, farming practices or storage, villagers are sceptical. There is initial disbelief and lack of seriousness to understand and adapt. But repeated attempts, patient explanations, proven track record of good results, capacity to easily mingle with villagers and above all a compassionate attitude will go a long way in winning farmers and their women folk. This kind of approach is not difficult for women. A sincere effort to convince villagers of the advantages of a new technology combined with the above approach always gives excellent results in adoption of technologies. Once the efforts bear fruit, farm women and little later men open up to female extension specialists. They become much more comfortable and feel bonded with women scientists. It is worth all the travails of travelling on rough and uneven roads to reach villages, put up with poor local conditions, which are anything but hygienic and lack facilities of food and washrooms. The affection of villagers over rides the discomfort of the effort of reaching villages and teaching mostly uneducated village folk.

CONCLUSIONS

It appears that the government needs to do much to address the problem of women agricultural scientists and agricultural extension specialists. At the same time, it is inevitable that women cannot dedicate to their scientific profession as much as men because of biological and sociological compulsions. Traits like care, warmth, compassion, patience and dedication make them highly ideal for breeding and extension education. Rural women need to be educated as to utilize the incentives that are in vogue. To remedy their unjust plight in spite of equal or more contribution to agricultural productivity and livestock maintenance, gender sensitization has to take place at all levels of administration and in societal attitude too.

References

Ahuja, A., Neeta Sharma & Shashi Tewari. 1999. Routine of farmwomen in hills - a reassessment. *Bhartiya Krishi Anusandhan Patrika.* 14 (1/2): 37-42.

Anon http://64.233.161.104/search?q-cache: 5chbNalZyVcJ: envfor. nic.in/ divisions/ic/ wssd/ doc

Balakrishnan, R. 2000. Widening gaps in technology development and technology transfer to support rural women. In: *Human resources in agricultural and rural development.* pp. 80-91.

Bamji, M. S. 2005. INSA examines Indian Women's access to and retention in science. *Curr. Sci.* 88(9): 1361-1362.

Chauhan, D. & P. Dayal. 2004. Assessment of physical strain among farm women. *Farm Sci. J.* 13 (2): 99-100.

DST. 1990 & 1996. R & D *statistics.* Department of Science & Technology, Government of India, New Delhi.

Geethalakshmi, G, K. P. Raghuprasad & S. V. Suresha. 2002. Comparative study on the farm activities performed by landless farm women labourers in the irrigated and rain fed areas of Shimoga District. *Curr. Res.* (Bangalore). 31 (5/6): 99-100.

Kar, S, A. K. Bandyopadhyay & B. Lahiri. 2002. Participation of rural women in agriculture and allied activities in some selected villages of Midnapore District of West Bengal. *J. Interacademica.* 6 (3): 400-405.

Ibid 1998b. Participation of farm family women in dairy and poultry enterprises. *J. Tropical Agri.* 36(1/2): 89-91.

Kolli, R. D. & M. C. S. Bantilan. 1998. Identifying indicators of gender-related impacts of improved agricultural technologies. Assessing joint research impact. *Proc. International Workshop on Joint Impact Assessment of NARS/ ICRISAT Technologies for the Semi Arid Tropics,* International Center for Research in Semi Arid Tropics, Patancheru, India.Dec. 2-4, 1996. pp.172-181.

Kumar, R. & R. M. Fulzele. 1998. Factors affecting the time utilization of farm women in household, agriculture and dairying activities. *J. Dairying, Foods and Home Sciences.* 17 (1): 55-59.

Kunwar, N. 2004. Decision making pattern among rural families of Western Uttar Pradesh. *Farm Sci. J.* 13 (2): 138-139.

Legg, H. G. 2002. Speech made during the third International Conference on women in agriculture, Madrid, Spain (http://www.wia.usda.goo.2002s peaches. html)

Menon, P. 2005. Far more women in science. *Frontline* 22(26): December, 17-30.

Mies, M. 1987. *Indian Women In Subsistence And Agricultural Labour.* Vistar, New Delhi.

Namboodiri, N. V. & R. L. Shiyani. 2001. Potential role of self-help groups in rural financial sector. *Indian J. agri. Econ.* 56 (3): 401-409.

Neelma Kunwar. 2004. Decision-making pattern among rural families of Western Uttar Pradesh. *Farm Science Journal.* 13 (2): 138-139.

Pal, B. C. & R. N. Seal. 2005. Bio-resource utilization by the Santals, Mundas and Oraons in the Hill Block, Dakshin Dinapur, West Bengal. *Environment and Ecology.* 23 (4): 935-941.

Puhazhenhi, V. 2000. *Evaluation study of self Help Groups in Tamil Nadu.* xi + 50.

Puhazhendhi, V. & K. J. S. Satyasai. 2002. Empowerment of rural women through self help groups–an Indian experience. National Bank News Review, Mumbai. 18 (2): 39-47.

Rama Rao, D. & Sandhya Shenoy. 1998. Professional women in agriculture. Final Report of DST Project, National Academy of Agricultural Research Management, Hyderabad.

Ramachandran, V. K., Madhura Swaminathan & Vikas Rawal. 2001. How have hired workers fared? A case study of women workers from an Indian village, 1997-1999. Working Paper, Centre for Development Studies, Trivandrum, India.

Reddy, P.G. & M.A. Rashid. 1997. A correlative analysis of characteristics of farm women with their farming performance in rice based farming system. *J. of Res.* ANGRAU 25(4): 60-62.

Roy, S. & G. Goyal. 1998. Perspectives of rural women in development programmes: a rationale analysis. Ecological agriculture and sustainable development: Volume 2 *Proceedings of International Conference on Ecological Agriculture:Towards Sustainable-Development*, Chandigarh, India, 15-17 November, 1997. pp 704-712.

Saikia. R. & L. DAS. 1998. Tribal women in agriculture. A study on involvement of Kabri women in jhum cultivation and their perceived drudgery in performance of farm operations. Indian. *J. Hill Farming*.11 (1/2): 13-18.

Sashidhar, P. V. K., B. S. Rao & M. S. Reddy. 2001. Time Utilization, suitability of month, duration and place of training programme for farm women. *J. Res.* ANGRAU 29 (2/3): 82-86.

Seth, A. & M. Bilgi. 2002. Report of the Ergonomics evaluation. Visit to the DFID-Gramin Vikas Trust, East India rain fed farming project, Silsoe Research Institute, Internal Report IDG/02/09.

Singh, R. K.P., K. N. Agarwal, K. K. Satpathy, S. Biswas, & M. Mawa. 2001. Analysis of drudgery prone activities and gender involvement in paddy cultivation. *Indian Hill Farming*. 14(2) : 61-67.

Singh, S. 2003. *Contract farming in India: impacts on Women and child workers.* Gatekeeper Series. International Institute for Environment and Development, Environmental Economics Programme.

Sisto, I. & X. Flores. 1999. Socioeconomic and Gender Analysis Programme (SEAGA). The training of trainer's workshop held in Malaga, Spain in June 1999. Agriculture and Environment for International Development. 93 (3/4): 163-169.

Sulaiman, R. V., J. Tahseen & M. S. Ashok. 2003. Cafeteria for Women in Agriculture. NCAP working paper 4. National Centre for Agricultural Policy, New Delhi.

Sureshkumar, R., N. P. K. Sushama & C. Bhaskaran. 1998a. Participation of farm family women in the agricultural operations and decision making in vegetable cultivation. *J. Tropical Agri.* 36 (1/2): 94-96.

Wedo. 1999. *Risks, Rights & Reform.* New York.

❑❑❑

SECTION-2
INTERACTIVE SESSION-FARM WOMEN AND SCIENTISTS

9

Role of Women in Indian Agriculture and its Allied Fields An Overview

PANDURANGAIAH, K & M.A. NARAYANA REDDY

ABSTRACT

Although women were not recognized as essential components of Agriculture for a long time, of late due recognition has been given to the role of women in agricultural production and steps have been taken to bring more and more women farmers into main stream, through Shree Shakthi Sanghas and Self Help Groups for women. Special projects aided by outside agencies and voluntary organizations such as WYTEP, IRDP, TRYSEM, DWCRA etc. provide training to farm women.These operations include selection and treatment of seed, protection of the seed or plant material, land preparation, application of manures and fertilizers, sowing or planting, nursery raising, irrigation, weeding, earthing up, pinching or thinning, training and pruning top dressing, plant protection, harvesting, transport and post-harvest operations such as threshing, winnowing, sieving, cleaning, drying, grading, packing and storage, joining hands with men etc. They also do marketing and disposal of the produce. For household use they take care in hulling, flowering, pounding etc.

In horticulture the varied activities carried by women comprise seed collection and extraction, raising seedlings/cuttings/sowing/transplanting/planting, manually applying fertilizers, stalking/tying/shading/mulching, pollination, bagging and thinning of flowers and fruits. Plant protection, harvesting, transporting, cleaning and grading, marketing, processing and preparation of value added products, mushroom cultivation and kitchen gardening including raising of medicinal and aromatic plants in the backyard are mostly carried out by women. Women are also adept at activities related to dairy, piggery, poultry, sericulture, fish culture and vermicomposting.

10

Women Oriented Training Programmes through Directorate of Extension, University of Agricultural Sciences, Bangalore

ESWARAPPA, G. K., P. RAGHUPRASAD
AND G. M. VARADARAJU

ABSTRACT

Women become major players in creating sustainable livelihood in rural areas. Though they are the major work force compared to men in agriculture and allied areas, they are considered as invisible entity. Their contribution in total labour force itself is about 55 to 60 per cent. They play multiple roles both in and outside the home by ensuring food security and substantiating nutritional security to the family as a whole. Empowering these women with knowledge and skill to take up vocational income generating activities is prime need of the hour. Further, building their capacities on managerial skills, decision making, and technical competency is also equally important area of concern.

In this backdrop, Directorate of Extension, University of Agricultural Sciences, Bangalore is extending specially designed capacity building activities to the rural women in various areas such as food processing, value addition, small scale agro based industries, crop and live stock production etc. To take up these activities 15 TOT centres are functioning under the Directorate of Extension including eleven Krishi Vigyan Kendras which are concentrating on empowerment of rural women. Along with this, Farmers' Training Institute and Bakery & Value Addition Centres located at Hebbal are mainly concentrating on capacity building of rural women. On an average, more than 5000 women are being trained every year through these institutions. As a result, many of them become successful entrepreneurs in areas like vermi-composting, mushroom production, dairy, floriculture, food processing etc. Additionally efforts are on to encourage group activities through women SHGs which are becoming lifeline of rural India.

Further, efforts are also initiated to involve women force in niche areas like environmental protection, natural resource management, energy conservation, rural agricultural waste management etc. to make them partners in development of the country.

SECTION-3
EMPOWERMENT OF WOMEN

11

Impact of Women Participation in Developing Vermi Village A Case Study in Karnataka

DHANANJAYA, B., R. VINAY KUMAR AND M.S.NATARAJU

ABSTRACT

One of the major distinctive agriculture extension concepts is the Intensive Agricultural Development Village concept or Village Adoption System, where in the Extension Guides who are working at *taluk* level select a suitable village and disseminate the need based and innovative technologies released by the University. In this connection Extension Education Unit of University of Agricultural Sciences at Mangalore has adopted Irvathur and Renjala villages of Karkala *taluk* in Udupi district and several educational activities like Field Trials, Result Demonstration, FLD, Block Demonstration and Integrated Farming System Demonstrations were carried out. Training programmes were conducted on Vermiculture, Dairy farming, Paddy cultivation, Cashew demonstration, IPM on Areca nut, Coconut etc., Besides establishing Milk Producer's Co-operative Societies in the villages, the key programme observed in the village was Vermiculture. To start with, Vermiculture was tried with one farm woman with 100 worms. Multiplication of worms and collection of casts and other Vermiculture techniques were demonstrated to the trainee farm woman. Later Vermiculture training programme was organized in the village and the technical information was given by the University Scientists, followed by the experience sharing of the demonstration farmer which influenced the other farm woman to adopt the technology in the villages. At present with the help of Shri Kshetra Dharmastala Rural Development Project-an NGO which is working in both the villages, thirty farm woman have adopted the Vermiculture technology and are producing vermicompost on small scale. On an average a single farm woman is having 2 tanks of 3x2x10 ft size and annually each farm women is getting Rs. 7500/. Totally 30 farm woman in these two villages have adopted the technology, as a result they are producing Rs. 1,87,500 worth of casts and earthworms from Irvathur and Renjala village, which are declared as vermi villages in the taluk and are model villages in Udupi district for Vermiculture technology.

INTRODUCTION

Vermi composting represents the growing branch of eco-technologies or green technologies that can enhance agricultural sustainability and secure agro-based livelihoods. By employing earthworms as versatile bioreactors, one can recycle organic wastes, stimulate rejuvenation of degraded land, conserve soil moisture, and improve crop yields. Integrated with other elements of agricultural extension activities, this can help villagers to build and run their own prospects around food sovereignty and livelihood security.

The University of Agricultural Sciences, Bangalore has an unique extension system in the country. Through extension education units, the concept of Intensive Agricultural Development Village (INTADV) is implemented wherein the Extension Guides of Extension Education Unit (EEU) of the University working at *taluk* level, select a suitable village and disseminate the need based and innovative technologies released by the university to the farming community. This paper narrates the success story of EEU, Mangalore that had adopted Irvathur and Renjala villages of Karkala *taluk* of Udupi district and developed them as model vermi villages.

Description of the village: Irvathur is a small village consisting of 240 farm families comprising of 35 big farmers, 200 small farmers, landless labourers and 5 other families. The village has facilities like four *Anganawadi kendras*, a middle school, two primary schools, one *Yuvaka mandala* for young men and one *Yuvathi mandala* for young women.

Majority of the farmers earn their livelihood through farming. Paddy is the major food crop followed by pulses, vegetables and other commercial crops. The village was adopted after a thorough discussion with local level development departments during the year 1999.

Extension approaches followed: The major extension activities carried out under the Intensive Agriculture Development Village programme during the year 1999 by the extension workers of EEU, Mangalore were as follows,

(1) *Demonstration and Field days:* Conducted several *Bhadra* and *Mukthi* paddy varietal demonstrations and organized field days. As a result, 70 per cent of the farmers in Irvathur village have adopted *Bhadra* paddy for *kharif* season and *Mukthi* for Rabi season. The farmers of the neighbouring village purchased 60 quintals of *Bhadra* and 28 quintals of *Mukthi* paddy variety seeds from the experimental village. The yield in the village has increased from 10 quintals to 18 quintals per acre (80.00% increase). This is also due to the application of self-produced vermicompost to the paddy fields.

(2) *Field trips / Field visits:* Around five Field trips / field visits were arranged involving the farmers to various vermicompost units and dairy units, by the Extension Education Unit, Mangalore to provide first hand information on success of new technologies.

(3) *Milk Producers Co-operative Society (MPCS):* A MPCS was started in the village during 1999. Earlier even to get 5 litres of milk farmers used to go to Karkala travelling 10 kms. After the initiation of MPCS in the village, every day 230 litres of milk is being sold to Karnataka Milk Federation (KMF). Twenty-five farmers/farm women are engaged in dairy production. In addition to this, farm yard manure (FYM) application to crops has increased and health condition of the farmers has also improved due to consumption of milk.

(4) Front Line Demonstrations on Black gram and Green gram were taken up in paddy fallow. In coastal Karnataka *kharif* paddy is completely rain fed. However, *Rabi* paddy is grown by using the seepage water, which flows for a period of 4 months after the completion of rainy days. Earlier farmers used to leave the land fallow after the harvest of second paddy crop. However, due to the introduction of Transfer of Technology (TOT) and Front Line Demonstration (FLD) in the village, farmers adopted fallow cultivation of black- and green gram. This involved an

expenditure of Rs. 6000 for land preparation and Rs. 8000 worth critical inputs for cultivation. This adoption yielded Rs. 50,400 worth Green gram and Black gram. After the demonstration, the farmers are following the technology till today, without any assistance.

(5) *Training Programme* : As high as 23 training programmes were conducted on Vermiculture, diary farming, paddy cultivation, cashew demonstration, IPM on areca nut, coconut etc. The major training programme was Vermiculture. To start with, Vermiculture was tried with one farmer with 100 worms. Multiplication of the worms and collection of casts and other vermiculture techniques were demonstrated to this farmer. Later Vermiculture training programme was organized in the village and the technical information was given by the University Scientists and followed by experience sharing of the demonstration farmer influenced the other farmers of the village to adopt the technology. At present with the help of Shri Kshetra Dharmasthala Rural Development Project (SKDRDP)-an NGO that is working in the village and with Karnataka State Department of Agriculture (KSDA), twenty six farmers have adopted the vermiculture technology and are producing vermicompost on small scale.

IMPACT OF THE TECHNOLOGY

Income generation: In densely populated, low-income farm families, making vermicompost is a successful method to process organic wastes. Small group of farm women in Irvathur village at Karkala *taluk* (30 farmers) have adopted the vermicompost technology on small scale. On an average, a single farmer is having two kgs of worms in tanks of 3x2x10 ft size. Annually each farmer is getting Rs. 7500 (Table 1).

Table 1. Income generated out of vermiculture enterprise by an individual farm woman

S. No.	Particulars	Income / farm woman/crop		Annual Income/ farm woman	
		Kg.	Rs.	Kg.	Rs.
1	Vermicompost	250	750	1500	4500.00
2	Worms	2	500	12	3000.00
3	Total	-	1250	-	7500.00

Note: Vermicompost sold @ Rs. 3/kg and Worms sold @ Rs. 250/kg

Table 2. Income generated out of vermiculture by the village (30 farmers)

S.No.	Particulars	I crop Income to the Village		Annual Income to the Village	
		Kg.	Rs.	Kg.	Rs.
1	Vermicompost	7500	22500.00	45000	1,35,000.00
2	Worms	60	15000.00	360	90,000
3	Total	-	37,500.00	-	2,25,000.00

Totally 30 marginal and small farmers in the village have adopted the technology, as a result of which they are producing Rs. 2,25,000-00 worth of casts and earthworms from the village, which is declared as vermi village in the *taluk* (Table 2). It is a model village in Udupi district for earthworms as well as vermiculture technology. On pooled basis all the farmers in the village together market the vermicompost and earthworms through the NGO, SKDRDP-Karkala to big and needy farmers in the district on cooperative basis.

2. *Impact on soil fertility*: Randomly selected vermicompost samples from Irvathur village were analyzed for plant nutrient contents. Vermicompost contained nitrogen, phosphorus and potash as well as micronutrients such as B, Cu, Fe, Mn, Mo, Zn, etc in sufficient quantities and in available farm. Besides beneficial micro organisms, micro flora, earthworm cocoons and some growth regulators are higher in vermicompost treated soils compared to normal soil. The details are delineated in Table-3.

Table 3. A comparative account of nutrient contents and microbial population of vermicompost and FYM

S.No.	Components	Vermi compost	FYM
1.	N (%)	1.8-2.0	0.30-0.60
2.	P (%)	0.6-0.9	0.15-0.40
3.	K (%)	1.0-1.6	0.30-0.80
4.	Zn (ppm)	84.0	14.50
5.	Cu (ppm)	61.5	42.80
6.	Mn (ppm)	509.7	69.00
7.	Fe (ppm)	1247.3	1465.00
8.	Fungi cfu/g x 10^5	8.4	5.8
9.	Bacteria cfu/g x 10^5	0.25	0.05
10.	Actinomycetes cfu/g x 10^7	1.3	0.21
11.	Azotobacter cfu/g x 10^4	10.2	2.45

Table 4. Participation of Farm women in Vermicompost Technology

S.No	Activities	Frequency	Percentage
1	Vermi bed preparation	26	86.66
2	Worm release	28	93.33
3	Slurry and moisture application	30	100.00
4	Ant and rodent management	30	100.00
5	Cast collection	30	100.00
6	Drying and bagging	27	90.00
7	Loading and Transportation	09	30.00
8	Marketing	15	50.00

It is clear from table-4 that, almost all the activities are carried out by farm women except the activities like loading, transportation and marketing which need male labour because the houses are scattered and far away from the main roads. Only under these circumstances male labour was required and rest of the activities was carried out by farm women only.

DISCUSSION

Some Multifunctional Attributes of Vermicompost: Vermicompost facilitates completion of the natural recycling loop of agriculture residue and other biodegradable wastes-a much desired objective of vermivillage. Vermicomposting is considered as an important tool for achieving this goal. The following are the multifunctional attributes of this technology, an understanding of which would help better integration of worm based innovations in vermivillage settings:

(1) *Simplicity and scale neutrality:* Vermicompost is easy to practice as an indoor as well as outdoor activity. Both farm and domestic wastes can be put to good use, it can be adopted on a smaller scale in one's own farm or at a commercial scale including large scale recycling units.

(2) *Plant nutrition:* Vermicompost as already mentioned contains nitrogen, phosphorus and potash as well as micronutrients such as B, Cu, Fe, Mn, Mo, Zn, etc in sufficient quantities and in available form, besides beneficial micro organisms, micro flora, earthworm cocoons and some growth regulators. As compared

to normal soil, vermicompost is five times richer in N, seven times in P, 11 times in K, two times in Mg and Ca, and seven times in actinomycetes (Bansal, 2002). However, vermicompost is a complex bio-fertilizer that must not be treated as a mere supplier of plant nutrients. In fact, the buffering action of vermicasts neutralizes soil pH, reduces soil toxicity, and makes minerals and trace elements readily available to plants. It also reduces leaching of nutrients, especially of N, from chemical fertilizers.

(3) *Efficiency of Conversion:* The vermicompost production capacity of earthworms can be gauged from the fact that each worm, weighing 0.50 to 1.20 grams, produces an equal quantity of compost in 24 hours. Thus, about 1000 earthworms can churn out about a kilogram of vermicompost per day (Sehgal, 2003). Composting through earthworms is completed in 50-60 days while anaerobic composting in pits takes 4-5 months.

(4) *Organic Input:* Vermicompost is a sweet smelling, fine, granular and stable organic matter. It is light weight, humus rich and richer than other types of composts. Importantly, vermicompost is a bio-input that can help farmers capture the growing market for green products and health foods. Regular use of vermicompost as part of an organic farming package leads to good yields. It also enhances quality, shelf life and nutritive value of horticultural crops. Organically grown products fetch a premium price in both domestic and international markets. Besides improving soil structure, these plants provide leaves (containing good amount of nitrogen) for making liquid fertilizer (vermiwash) as well as for mulching.

(5) *Complementary effects:* Worms multiply fast. Under optimum conditions, eight worms can produce 1500 new worms within six months. The mature worms, high in protein, are often sold as fish bait or used directly to supplement animal feed. The earthworm biomass can also be processed to produce vermimeal for use as a protein rich livestock feed and fish feed.

SUGGESTED STRATEGIES FOR ADOPTION OF VERMICOMPOTSING

(1) *Creating demonstration effects*: A successful commercial venture, based on regenerative principles, can result in a demonstration

effect convincing people about the merits of sustainable practices. To be considered as a viable alternative, large-scale vermicompost units must be ecologically and commercially sustainable. Setting up of commercial scale vermicompost production plants, which run on a competitive basis without subsidy, will enhance the credibility factor. Production of compost through industrial plants utilizing earthworms and other biological converters along with extensive use of green manure crops in agricultural lands results in sustainable soil fertility.

(2) *Multi Level capacity building*: Vermicompost technology needs to be convincingly presented as a healthy alternative to chemical based agriculture. Building awareness about regenerative practices and dissemination of successful experiments will facilitate such transition. Training of farm women, grassroots organizations and extension agencies in vermicompost production methods, and inclusion of the subject in the agricultural education curriculum will help capacity building. Equipping housewives and home gardeners with appropriate skills to produce their own vermicompost from domestic waste can be a very useful intervention. Sensitising policymakers to the significance of regenerative food production methods will go a long way in carrying forward the gains of the current experiences.

(3) *Adherence to quality standards*: Establishing appropriate quality norms for vermi compost will ensure consumer satisfaction and thereby continuity of its utilization. Care is needed in separating the organic matter from the waste stream before using as feedback in order to avoid heavy metal accumulation in earthworm tissue.

(4) *Fine tuning extension messages*: Besides highlighting its plant nutrient value, greater emphasis needs to be given to other desirable attributes of vermicompost to gain confidence of the farm women. A detailed economic analysis will help in understanding the viability of this technology better.

(5) *Development of effective extension materials*: For speedy dissemination of this important technology, a variety of extension and communication aids may be produced. Efforts are also needed to develop packages of IT based user-friendly materials to support easy access to relevant information to farm women.

References

Bansal, G.L. 2002. Vermicompost: Potential livelihood for rural poor and a source of nutrients in agriculture. *Paper presented in workshop on Bio-diversity Conservation* (12-14 Aug. 2002). National Institute of Rural Development, Hyderabad, India.

Sehgal, R. 2003. Stories of change: Earthworms to the rescue. *Info Change News and Features (Sept. issue).*

12

Participation of Farm Women in Jasmine Production

DHANANJAYA, B., R. VINAY KUMAR
AND M.S. NATARAJU

ABSTRACT

Jasmine production in coastal Karnataka was confined to only few farmers with some specified pockets of coasted belt. Only a hand full of farm women in Belman, Manchakal areas were cultivating jasmine successively. The scientists of Extension Education Unit (EEU), Mangalore Campus of University of Agricultural Sciences, Bangalore visited the area and conducted a Survey during September 2001 and gathered information about farmers practicing jasmine production in the area and came out with some success stories from the jasmine growing belt. The jasmine market was studied especially the demand for jasmine flowers (*Udupi Mallige*). It was observed that there was huge demand for *Udupi Mallige*. The extension specialists conducted training programmes in different locations in the coastal belt of Karnataka which are not traditional jasmine cultivated areas. As a result the technology has spread to the entire coastal belt. At present farm women are supporting themselves soley by cultivating jasmine in an area of 3-5 cents with a minimum of 20 bushes. Further, the KVK's which are working in Udupi and Dakshina Kannada districts are also working in this direction through On Farm Testing, Frontline Demonstration, Off campus and On campus training. Since it involves only women labourers, it is one of the most important and viable subsidiary as well as main occupation of the women folk in the district.

INTRODUCTION

Jasmine (*Udupi mallige*) production in coastal Karnataka has gained momentum in recent years especially amongst women having very small land holdings. Earlier jasmine production was confined to only few farm women in some specified pockets of coastal belt. A hand full of farmers in Shankarapura and Manchakal areas were cultivating jasmine successfully and it was not practiced in other parts of the coastal belt. The scientists of Extension Education Unit (Mangalore) of University of Agriculture Sciences, Bangalore conducted a survey during September 2000 and gathered information about farmer's practice of Jasmine cultivation and came out with some success stories from the jasmine growing belt. Scientists also studied the Jasmine market. It was observed that there was huge demand for *Udupi Mallige* (Jasmine flowers) in local as well as in far off places like Mumbai. Further, scientists of Extension Education Unit (EEU), Mangalore have also demonstrated the proper technology of jasmine cultivation to the farmers. In this context the present study was undertaken with the following objectives:

1. To study the participation of women in jasmine cultivation.
2. To study the socio-economic profile of jasmine growers.
3. To find out the constraints faced by jasmine growers.

METHODOLOGY

The study was conducted in selected *taluks* of Udupi and Dakhsina Kannada districts of coastal Karnataka. Karkala, Puttur, Udupi and Mangalore *taluks* were selected for the study by the extension guides of University of Agricultural Sciences, Mangalore campus. In each *taluk*, ten-selected farm women were provided with good quality Jasmine saplings and the Extension Guides of the university working at grass root level provided technical know-how. Thus, 40 selected farm women were encouraged to grow jasmine scientifically. Data were collected by personal interviews through structured interview schedule. The collected data were analysed using frequency and percentages.

RESULTS AND DISCUSSION

1. *Socio-Economic Profile of Jasmine Growers*: The detailed information on socio-economic profile of Jasmine growers is provided in Table 1 and is summarized as below:

Age: It was observed that about 67.50 per cent of the respondents were from young age group followed by 25.00 per cent in middle age group. The old age group comprised of only 7.50 per cent. Since jasmine cultivation involves deftness especially for fast plucking of Jasmine flowers in the early morning and quick preparation of garlands, it suits women of younger age.

Gender: Maximum number (95.00%) of the respondents who grew jasmine were female, while least number of respondents (05.00%) were male. It indicates that jasmine cultivation needs more of female labour than the male labour because, women labourers are better suitable as they are quick in plucking flowers and making garlands.

Caste: Table-1 further indicates that majority of the Jasmine growers (72.50%) belong to backward caste followed by forward caste (22.50) and least number of respondents belonged to scheduled caste (5.00%).

Occupation: It was observed that, majority (65.00%) of the respondents were involved in farming & other activities followed by people involved in farming only (25.00). It may be due to the fact that, jasmine cultivators are small and marginal farmers and as such they cannot depend only on jasmine cultivation. Apart from cultivating jasmine, they are also involved in *Beedi* making.

Land Holding Pattern: Findings of the present study indicated that, majority of the respondents had marginal (45.00%) and small (37.50%) land holdings but none of the large farmers cultivated jasmine in the coastal belt because it involves huge amount of women labour for plucking flowers and preparation of garlands.

House Type: Large majority of the respondents had tiled roof house (90.00%) and least number of the respondents had terraced house (7.50 %) and thatched roof (2.50 %). Since majority of the respondents were literates, they might have approached the *gram panchayat* and utilized the benefits of Housing Schemes for rural folk being implemented by the government of Karnataka and constructed tiled houses to withstand the heavy rainfall of the coastal area.

Social participation: It was noticed that, nearly half of the respondents (52.50%) had medium level social participation and were participating in youth clubs, group meetings and *Grama Panchayats*. Only 10% were socially active while 37.50% were socially not very active.

Mass media participation: Majority (47.50%) of the respondents had medium level of mass media participation followed by low mass media participation. Majority of the respondents subscribed newspapers, owned radio and television sets, listened to agricultural programmes and read information related to agriculture. High media participation was seen in least numbers (17.50%).

Table 1. Socio-economic profile of jasmine growers of coastal Karnataka (N=40)

S. No.	Variables	Classification	Frequency	Percentage
1	Age	Young (<34 years)	27	67.50
		Middle (34-59 years)	10	25.00
		Old (>59 years)	03	07.50
2	Gender	Female	38	95.00
		Male	02	05.00
3	Caste	Forward caste	09	22.50
		Backward caste	29	72.50
		Scheduled caste	02	05.00
4	Occupation	Labour	04	10.00
		Farming	10	25.00
		Farming and other	26	65.00
5	Land holding pattern	(a) Landless	02	05.00
		Marginal	18	45.00
		(b) Small	15	37.50
		Medium	05	12.50
		Large	00	00.00
6	Education	Illiterate	00	00.00
		Functional literate	04	10.00
		Primary School	12	30.00
		Secondary School	15	37.50
		Collegiate	09	22.50
7	House type	Thatched	01	02.50
		Tiled roof	36	90.00
		Terraced	03	07.50

Contd...

Contd...

8	Social participation	Low	15	37.50
		Medium	21	52.50
		High	04	10.00
9	Mass media participation	Low	14	35.00
		Medium	19	47.50
		High	07	17.50

1. *Income obtained from Jasmine Production:* It is evident from the Table 2 that, majority of the respondents belonged to medium level of income category mainly generated from jasmine cultivation (Rs.4000/- month). This level of income was possible with supplemental earnings from diary and agricultural waste. This was made possible by the guidance given by scientists to farm women regarding jasmine cultivation. This motivated small and marginal farm women to adopt jasmine cultivation. The scope for improving jasmine yield is great, because twenty percent of farmers are still getting lower yields. With proper guidence and motivation from scientists, these farmers can also increase the yields of jasmine in their plots. It was surprising to note lowest response from farmers who earned maximum income of Rs.6000/ month from jasmine cultivation.

Table 2. Income generation from jasmine cultivation

S. No.	Category	No. of respondents	Average Monthly Income (Rs.)
1	Low	10	2000.00
2	Medium	22	4000.00
3	High	08	6000.00

2. *Constraints Faced* by the jasmine growers are listed in table 3. Among the enlisted constraints, low yield in winter season (92.00%) was the major constraint experienced by farmers followed by pest and disease attack (88.00%), non availability of recommended chemicals (80.00%) and work pressure (72.00%).

More than the half of the respondents (56.00%) expressed labour required to pluck flowers and preparation of jasmine garlands was a major impediment, followed by financial shortage (44.00%). Marketing was considered a problem only by 12% respondents.

Table 3. Constraints faced by the jasmine growers

Constraints	Frequency	Percentage
Low yield in winter season	23	92.00
Pest and disease attack	22	88.00
Non-availability of recommended chemicals	20	80.00
Work pressure	18	72.00
Labour required to pluck flowers and preparation of jasmine garland	14	56.00
Financial constraint	11	44.00
Marketing problem	03	12.00

CONCLUSION

The technology of jasmine production has spread to the entire coastal belt of Karnataka, which was confined earlier to only certain pockets of the area. At present very small and marginal farm women are deriving their entire livlihood from an area of 3-5 cents with a minimum of 20 bushes. Scientists are advising the farm women to have 25-30 bushes each in addition to a main occupation, which should give additional and daily income to the stakeholders. The expansion in jasmine cultivation and marketing were due to the study on the demand and price of jasmine carried out by the Extension Education Unit. This has resulted in obtaining sustained income. Further, the Krishi Vignana Kendras which are working in Udupi and Dakshina Kannada districts are also working in this direction to give fruitful technology to the small and marginal farm women through On Farm Testing, Front Line Demonstration, off campus and on campus training programmes.

13

Women Empowerment As A Challenge for Agricultural Development in India

BHARATHAMMA. G. U.
B. K. NARAYANASWAMY AND A. ASHWINI

INTRODUCTION

In the history of human development, woman has been as important as man. In fact, the status, employment and work performed by women in society is the indicator of a nation's overall progress. Without the participation of women in national activities, the social, economical or political progress of a country will stagnate. Empowerment is a multi-dimensional process, which should enable women to realize their full identity and power in all spheres of life. It consists of greater access to knowledge and resources, greater autonomy in decision making to enable them to have greater ability to plan their lives, or to have greater control over the circumstances that influence their lives and free from shocks imposed on them by customs, beliefs and practices. Generally development with justice is expected to generate the forces that lead to empowerment of various sections of women and to raise their status. There are certain operations in agriculture in which women dominate in production, post harvest management and agro processing. Various agro based activities for women entrepreneurship have been identified and initiated. It is believed that once we are able to instill confidence in the rural women entrepreneurs for making the quality produce, the rural scenario will be changed beyond imagination. Different approaches are needed to realize agricultural development in view of the above.

(1) *Reflect as an Approach for Agriculture Development* : Rapid Rural Appraisal (RRA) is one of the approaches for diagnosing problems of rural women for their agriculture development. This requires a systems approach where extension education efforts have to be combined with argilultural support activities to create a favorable climate. RRA is a quick approach to assess the needs of rural women in villages by a multidisciplinary team through analyzing the situation and involving creative and structural use of investigation tools in a more accurate and cost effective manner. However, Regenerated, Freiren, Literacy Through Empowering Community Techniques (REFLECT) is a new approach to adult literacy which fuses the theory of Paulo Freire and the practice of Participatory Rural Appraisal (PRA). The approach is more effective at teaching people to read and write and much more effective at linking literacy to agriculture development. In this approach literacy circle develops its own learning materials through the construction of map, matrisis, calendars and diagrams that represent local reality. It systematizes the existing knowledge of participants and promotes detailed analysis of local issues. In the REFLECT, the graphic is initially constructed on the ground using locally available materials. It is then transferred to a large sheet of paper or card using simple pictures. The approach is highly participatory and depends on inputs from each participant's experience to construct the basic learning materials. Once the graphic is completed, focused questions can explore key issues. These questions are always related to participant's life and his community. The frame work for graphic is produced by villagers themselves. Hence, the participants feel the issues are theirs not some one else's and as a result the discussions are much more likely to lead to the identification of local solution and local actions. Reflect circles can act as a catalyst for agriculture development.

(2) *Demonstrations as an Approach for Agriculture Development (Figure 1):* The era of globalization has provided enormous

opportunities for agriculture development. Inspite of constitutional rights of equality for all without any discrimination, there is need for suitable steps for creating favorable environment to enlist rural women's participation for agriculture development. There is need to create strong desire among rural women with clear goals for agriculture development. Demonstrations have proved to be very essential to convince rural women on the benefits of scientific agriculture. Demonstrations are used in various areas like agriculture production, processing and storage, involving women in technology development, technology dissemination, and technology utilization. Narayana's Lotus Model explains seven important aspects of demonstrations namely (1) More senses are involved in the techniques (2) Most credible among extension methods (3) All extension teaching methods can be used (4) All types of clientele groups can be involved (5) It is a leadership development method (6) It is a horizontal teaching technique (7) It will have direct effect on behavior.

NARAYANA'S LOTUS MODEL TO SHOW ROLE OF METHOD DEMONSTRATION

(i) More senses are involved.

(ii) Most credible methods.

(iii) Individuals, group and mass methods.

(iv) Helps for Development of abilities and standards.

(v) The results can be spread horizontally.

(vi) All type of Clientale can be involved.

(vii) It will have direct effect on behaviour.

Fig. 1. Seven orchestrated, concerted, comprehensibles of demonstration are shown on the lotus for women empowerment.

Modern-day economists speak more and more in favor of investment on rural women for agriculture development in developing countries. Though this may sound like a paradox, this is the basic aspect of developmental plans because it works both ways. If people are not aware of benefits of agricultural schemes all their contributions will be uninspired. So the tempo of awareness and involvement should be increased. Village women, being traditionally innovative, make practical and wise combination and judicious use of agricultural practices. Thus it is necessary to formulate action plans for involving women in agriculture and take simple and practical technologies to them for developing required competence so that they can overcome the hardship arising from unemployment, under employment, uncertain employment etc.

(3) *Effective Communication as an Approach for Agriculture Development (Fig.2):* Empirical studies have proven that economic growth in advanced countries is due to human skills rather than to capital. Further, investment on rural women has directly contributed to economic growth through promotion of knowledge, application of science and technology to the production process. Research, training, development of right type of attitudes, values, interest etc. are conducive to higher output of agriculture development. However, for lack of effective communication many modern agricultural technologies are not reaching the rural women. In addition much misunderstanding results from faulty communication. Thus there is need for effective communication for promotion of modern agricultural technologies among the rural women. Good communication should consist of creating understanding, imparting of knowledge and help rural women to gain a clear view of agricultural technologies. It is the responsibility of those involved in promotion of modern agriculture technologies to familiarize themselves with seven attributes of communication to be effective as explained in Narayana's Wheel Model i.e. (1) Credibility: A Climate of belief and earnest desire. (2) Context: Realities of situation must provide for participation and playback. (3) Content must have meaning for rural women. (4) Clarity: The message must be clear and simple; words must mean the same thing to rural women as they do the promoter. (5) Channels: Established channels of communication for rural women must be effectively used. (6) Consistency: Communication is an unending process. It requires repetition to achieve penetration; it should be consistent. (7) Capability: This refers to inherent capacity to learn. The above formula ensures good and effective communication to transform rural women in to confidant women.

(4) *Competency development as an approach for agriculture development (Fig.3) :* The optimum utilization of natural resources will be

possible only when we have skilled and efficient manpower. Therefore, there is an urgent need to impart the necessary skills, attitude and knowledge to rural women for optimum utilization of natural resources. Thus, there is also a need to mobilize the women resources with the purpose of enabling them to participate effectively in the task of agricultural development. Unless the women resources are adequately trained and developed to the required levels of proficiency and requirement, they cannot be deemed to be effectively managed. Narayana's Lotus Model explains four major aspects of rural women with seven sub items under each head for agriculture development like (1) Knowledge (Subject matter, Farming systems, Input and Marketing, Farm and farm families, Farmer's problem solving, Farm economics and Extension methods) (2) Attitudes (Serving clients, Technological know-how, Field work, Organization, Farmer's interest, Problem solving, Self development) (3) Skills (Communication, Technological Training, Diagnostic Solutions, Organizational, Demonstration) (4) Attributes (Self confidence, Personality, Conduct, Empathy, Experience, Flexibility and Self development).

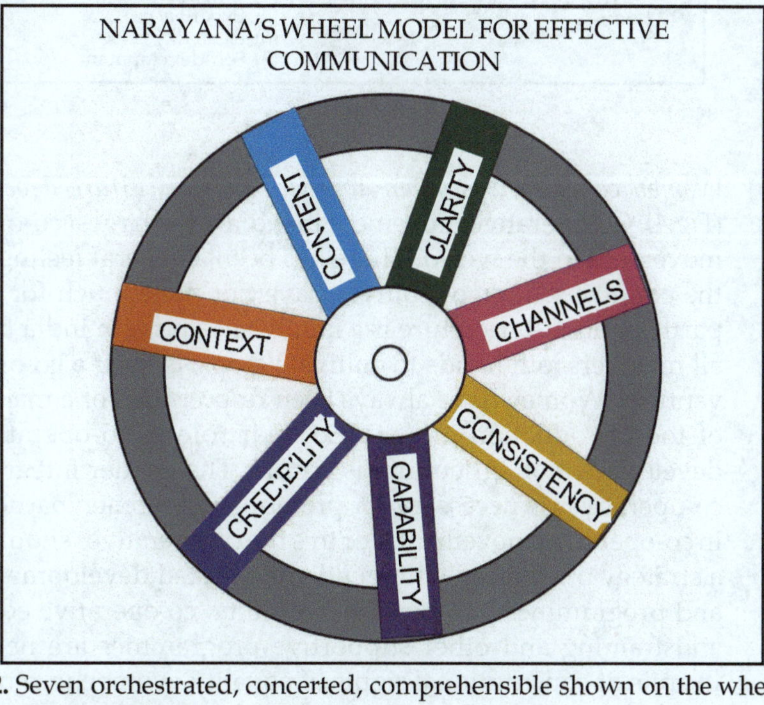

Fig. 2. Seven orchestrated, concerted, comprehensible shown on the wheel for effective communication

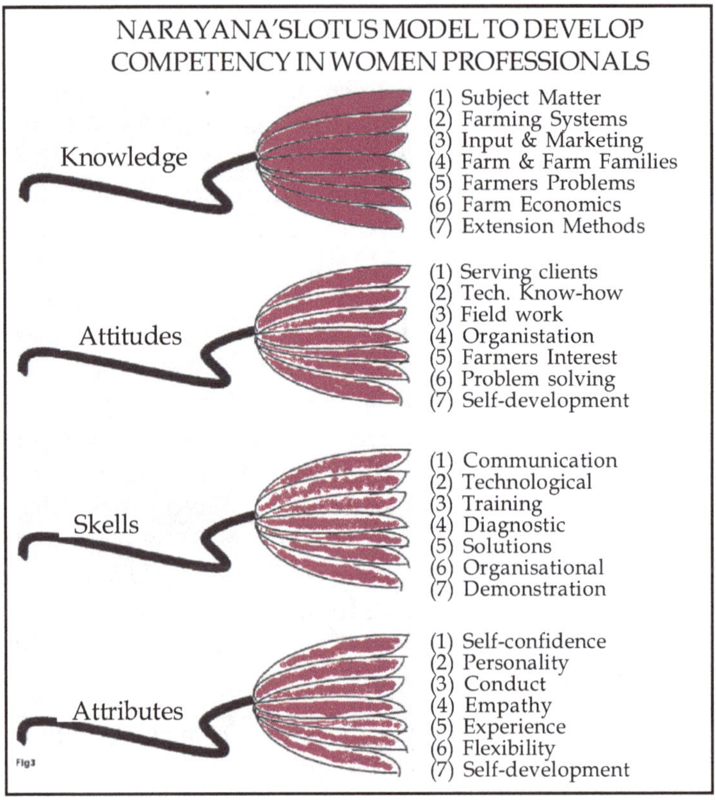

Fig. 3.

5) *Women co-operatives as an approach for agriculture development (Fig.4):* Co-operative movement in India is the largest co-operative movement in the world. There has been a general feeling within the country that co-operatives have not done much for women participation. Agriculture is a family enterprise in India in which all members join hands to cultivate crops, is itself a co-operative venture. Women have always been responsible for a major share of food production in Asia. So their role in co-operatives for development of agriculture is crucial. The gender integration in co-operatives is necessary for promotion of women participation in co-operative movement. For this the co-operatives should adopt a strategy of equality with gender integrated development plans and programmes. For women's equality, co-operative education and training and other supportive programmes are necessary. In spite of constitutional rights of equality, women in most cases cannot avail opportunities. It is clear that there is an urgent need

for initiating crash programmes for upgrading the status of women. Further, social status can be improved with financial support. The best instrument to achieve this is Mahila Co-operatives whose main objective is to generate the idea of doing things by women themselves in a co-operative and mutually benefical way. Narayana's Wheel Model explains seven basic steps to be followed by women co-operatives for agriculture development i.e. (1) Getting the attention of rural women towards choosing best agriculture technology. (2) Stimulating rural women's interest to develop intelligence. (3) Arousing the rural women's desire for agriculture information. (4) Convincing the rural women to take up agricultural production according to market requirement. (5) Getting the action by going down and looking to the needs and interest of rural women. (6) Follow up action (7) Finally, it is very essential to see that rural women should have a sense of satisfaction. Growth of co-operatives among women is one positive indicator of economic well being of any country.

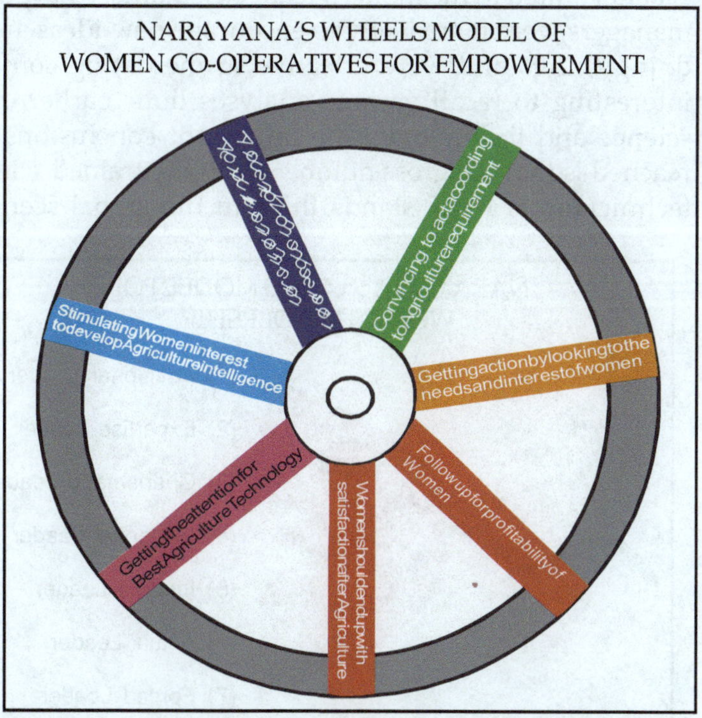

Fig. 4. Seven orchestrated, concerted, comprehensible Techniques for women empowerment are shown on the wheel

(6) *Leadership development as an approach for agriculture development* (Fig.5): The changing role of the Government, matching supply and demand in agricultural units requires development of women staff as the largest number of working women in India are engaged in farming operations either as cultivators or as agricultural labourers. Most of the science and technology programmes in developing countries have failed so far because they haven't taken into account properly the gendered nature of development. Women have entered the professions in unprecedented number in what they believe to be the first step, towards the leadership. Narayana's Lotus Model explains seven leadership styles namely (1) Positional leader (2) Expertise leader (3) Charismatic leader (4) Influential leader (5) Implied leader (6) Actual leader and (7) Forced leader. Although women constitute nearly 50 per cent of the population, their participation in economic activities is approximately 34 per cent. A study conducted by a Sydney based consultancy firm, DDI Asia pacific indicates that women work harder and their feminine approach is better suited to the modern team based office. Further, women managers were more likely to encourage new ideas and ways of doing things and were effective in mediating conflicts. It is interesting to recall various analyses done earlier on gender, science and technology. The important conclusions they had reached is that the gross number of highly trained scientists and technicians in India stands third in the global scene, but the

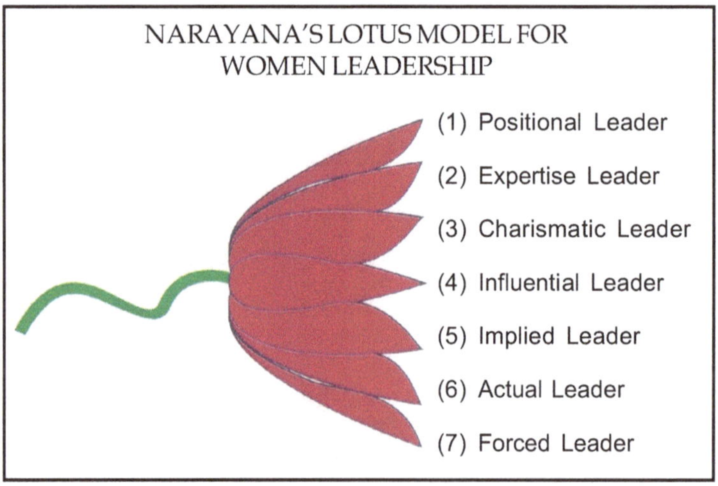

Fig. 5. Seven orchestrated, comprehensible Qualities shown on the lotus

representation of women in science subject is very low and representation in technology transfer is very negligible. Hence, Extension services need to provide appropriate technologies to rural women and the best strategy for reaching them is by using female extension agents.

(7) *Effective management as an approach for agriculture development* (Fig. 6): The empowerment of women is hypothesized as an important outcome of their high level participation in agriculture development programmes. Empowerment is a leading out come of successful capacity building at individual and Institutional levels. An empowered woman is one who can take initiative, exert leadership, display confidence, solve new problems, mobilize resources and undertake new actions. The larger perspective of empowerment is the process beginning from input procurement in agriculture to processing and marketing and the skills gained at each step. Hence, options available to women need to be widened. The rapid agricultural development is intrinsically linked to the social, economic and political empowerment of women. Ensuring gender equality by removing gender discrimination is the key to real empowerment. Power means many faces, at the individual level i.e., personal power, feeling in control of one's own environment and feeling good about one's self. Interpersonal power is the ability to influence others and organizational power is the ability to mobilize

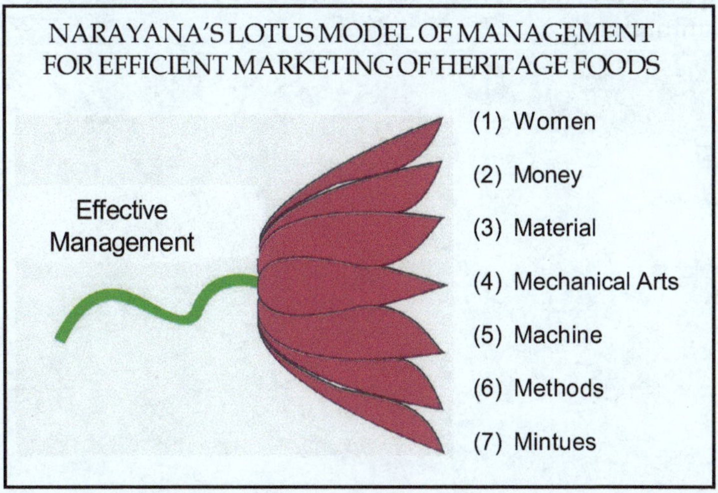

Fig. 6. Seven orchestrated, comprehensible key elements are shown on the Lotus

resources, the ability to get things done. By empowering, women will become best managers of agriculture. Narayana's Lotus Model explains seven inputs for effective management i.e. Women, Money, Material, Machinery, Methods, Mechanical arts and Minits.

CONCLUSION

Farm women are the back bone of Indian agriculture. The rapid agricultural development is intrinsically linked to the social, political, legal, cultural and economic lives of rural women. By empowering, women will become best managers of agriculture. Empowerment comes from Women's groups who seek to empower themselves through greater self-reliance. They have right to determine their choices in life. They also try to gain control and access to resources. Village women are quite innovative and also have indigenous knowledge, There is need to create strong desire among rural women with clear goals for agriculture development. Further, demonstrations are reported to be very essential to convince rural women to adopt scientific agriculture. Hence, demonstrations are used in various areas like Agriculture production, processing and storage, in technology development, technology dissemination, and technology utilization with appropriate policy for agriculture development. Most of the development specialists recognize that women play a vital role in dissemination of agricultural technologies. The process of agriculture development would be incomplete and lap sided unless women are fully involved in marketing dissemination.

14

Poverty Alleviation Programmes and Women The Karnataka Scenario

PUSHPA, B

ABSTRACT

Various studies put the percent of rural women engaged in agriculture and related activities in India around 65-70%. A majority of them are from economically and socially deprived category and qualify as "poor" or below poverty line. In rural India women play the role of agricultural labourer, fuel gatherer, food provider and home maker. Both central and state governments have framed and implemented various programmes to improve the living conditions of rural people. Measures have also been taken to provide employment and alleviate poverty. Generally these programmes fall into the category of wage employment, self-employment, housing and habitat development, and watershed development programmes. The last mentioned focuses on various water supply schemes to the rural poor and socially deprived. These programmes are evaluated in this paper. Although they are intended to benefit all the poor and jobless in rural Karnataka, the benefits reach women too because of their role as homemakers and food providers. Be it wage earned or drinking water or shelter, these programmes, if implemented sincerely, make the life of farm women less laborious and more comfortable. The programmes are explained in detail so that women agricultural scientists and extension specialists become well aware of the opportunities available to rural women and will be in a state of awareness if sought for guidance and problem solving. Such aknowledge can make them better teachers and counsellors at village level.

INTRODUCTION

India has a long history of Government programmes for poverty alleviation. They are national rural employment programmes like Jawahar Rozgar Yojana (JRY), target-group specific programmes like Integrated Rural Development Programme (IRDP), Small Farmers Development Agency (SFDA), Training of Rural Youth for Self Employment (TRYSEM), Jawahar Gram Samriddi Yojana (JGSY), Employment Assurance Scheme (EAS), Sampoorna Grameena Rozgar Yojana (SGRY), Swarna Jayanthi Gram Swarozgar Yojana (SGSY), Indira Awaas Yojana (IAY), Pradhana Mantri Gramodaya Yojana (PMGY), Grameena Awaas Yojana (PMGY), Drought Prone Areas Programme (DPAP), Desert Development Programme (DDP), Bore wells with Hand pumps Scheme, Mini Water Supply Scheme, Piped Water Supply Scheme, Pradhan Mantri Gram Sadak Yojana (PMGY), Nirmal Grama Yojana, Rajeev Gandhi Housing Corporation Limited (RGHCL), Ashraya Scheme (Rural), Rural Ambedkar, Special Area Programmes and Food Subsidy.These Government programmes encourage both self-employment and wage-paid employment and income generating opportunities in addition to making food available at subsidized prices under Public Distribution System (PDS). Thus, it was and is expected that prior, absolute rural poverty and this spending are inversely associated (i.e., <0.01). In other words, as Government Expenditure on Poverty Programmes (GEPP) increases rural poverty ratio diminishes.

Area development programmes can be classified into the following three broad categories.

1. *Wage Employment Programmes* (Sampoorna Grameena Rozgar Yojana -I Stream and II Stream)
2. *Housing and Habitat Development Programmes* (Indira Awaas Yojana - Construction and Up gradation of Houses, Pradhana Mantri Gramodaya Yojana - Grameena Awaas Yojana and SWASTHI)
3. *Watershed Development Programmes* include DPAP (Drought Prone Area Programme); DDP (Desert Development Programme); WGDP (Western Ghats Development Programme); IWDP (Integrated Wasteland Development Programme)

The department of Rural Development and Panchayat Raj is implementing a number of programmes for poverty alleviation; they are :

1.1 Wage Employment Programme

Until 2001-02 the Rural poverty alleviation programmes like JGSY and EAS provided wage employment to the rural poor by engaging them in the creation of economically productive and socially beneficial assets like roads, school buildings, irrigation wells, anganwadi buildings, community halls and land development works.

1.2 Jawahar Gram Samridhi Yojana

In order to mitigate rural un-employment and underemployment, the Government of India introduced the Jawahar Rozgar Yojana (JRY) on 1.4.1999 by merging the then existing wage employment programmes of NREP and RLEGP. The main objectives of the programmes were:

a) Generation of additional gainful employment for un-employed and under-employed men and women in rural areas.

b) Creation of durable community assets.

This programme is a centrally sponsored scheme and the expenditure under this scheme is funded by the centre and state in the ratio of 80: 20. *Zilla panchayats* and *grama panchayats* are responsible for the implementation of this programme. The funds are allocated in the ratio of 15 : 15 : 70 to the *zilla panchayats* (ZP), *taluk panchayats* (TP) and *gram panchayat* (GP) respectively. Works like social forestry, soil and water conservation, water harvesting structures, community irrigation wells, construction of drinking water tanks, farm ponds, drainage, sanitary latrines and other rural sanitary programmes, formation of roads, construction of *anganawadi* and primary school buildings, *grama panchayat* office and other community buildings, etc., are taken up under this programme. There is no sectoral earmarking of funds but for 22.5 per cent of the total allocation is for the benefit of SC/ST. The *Gram panchayat, taluk panchayat* and *zilla panchayats* are free to select the works depending on their local needs.

The ratio of the wage and non-wage component for all the works is maintained on a 60: 40 basis. Food grains are distributed at 2 kg per man-day as part of wages at subsidized rates. This is optional. The minimum wages are Rs.26/- per day. From the inception of this programme, 3411 lakh man-days of employment have been generated incurring an expenditure of Rs.1156.26 crores.

Later this programme was renamed as Jawahar Gram Samridhi Yojana which operated until 2001-02. Jawahar Grama Samridhi Yojana (JGSY) has been successful in providing employment to the rural poor and in the creation of durable community assets. From the inception of this programme in 1989 till March 2002, 4333 lakh man-days of wage employment were generated. The total expenditure during this period was 1685.49 crores. Fig. 1 explains, total expenditure and man-days of employment generated till the end of March 2002 under this programme. In the year 2001-02 Jawahar Gram Samridhi Yojana (JGSY) and Employment Assurance Scheme (EAS) were merged to become Sampoorna Grameena Rozgar Yojana (SGRY).

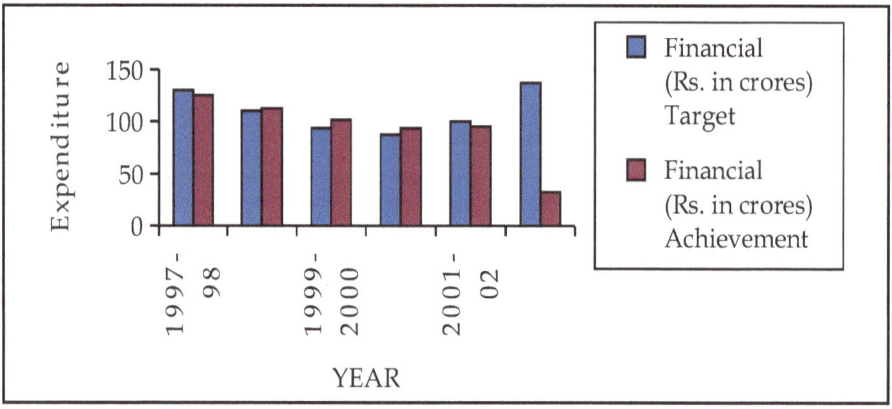

From ANONa (1997-2002)

Fig 1. Total expenditure and Man days of employment under JGSY

1.3 Employment Assurance Scheme (EAS)

The EAS came into existence from 2nd October 1993, as JRY alone could not provide sufficient employment opportunities to the rural poor. Later on, JRY was intensified by adding new blocks. Thus the number of blocks were increased from 94 to 162 by the end of 1996-97. Since 1997-98 all the 175 blocks in the state are covered under employment assurance scheme. The purpose of the scheme is:

a) To provide assured employment to those who are in need for work during the lean season.

b) To take care of community assets.

Under the scheme, at least 100 days of employment is assured to both men and women between the age group of 18 to 60 years. Works can be started anywhere in rural areas, where people are ready to work even in very small numbers.

Hitherto, a demand driven scheme, the employment assurance scheme has metamorphosed into an allocation-based scheme with effect from 1.4.1999. The watershed component, which accounted for a major chunk of funds released under EAS, has been taken out of EAS. Seventy per cent of the allocation was earmarked for the taluk panchayats and 30 per cent was retained at the district level for taking up works of emergent nature. Fifteen per cent of the funds was provided for maintenance of the assets created under the said programme. Works taken up under EAS were targeted to be completed within a period of two years. The state Government gave additional instructions to the zilla panchayats with a view to streamline the implementation of the scheme and to ensure the optimum utilization of resources. According to these guidelines not less than 25 per cent of the allocation was to be spent for *anganawadi* and primary school buildings. Minimum 25 per cent of the funds was earmarked for construction of ground water recharge structures especially around drinking water sources. All the 175 blocks of the state were covered under the EAS. The scheme aimed to provide wage employment to the rural poor in addition to the creation of community assets. The financial and physical progress during 1997-2002 is given in Fig 2.

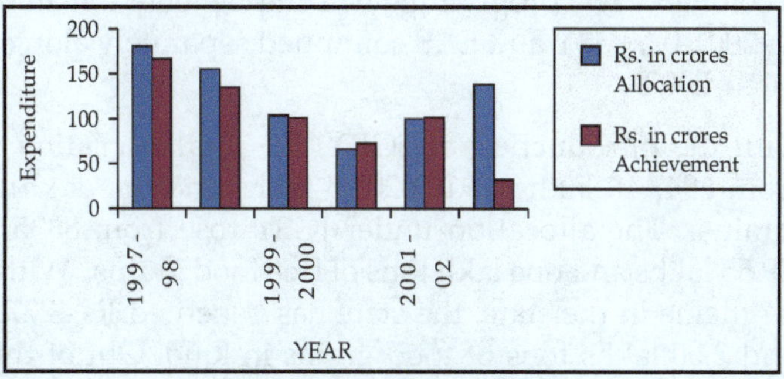

From ANONa (1997-2002)

Fig 2. Financial progress under EAS

Fig. 2 shows that there was a gap between the total funds available under this programme and the achievement. During 1997-2002, the achievement was lagging behind compared to the total funds available. Further the physical achievements in terms of lakh man-days also decreased from 352.76 during the year 1997-98 to 142.40 man-days during the year 2001-02. Even from the point of food grains distribution, it was very less compared to the total funds available.

1.4 Sampoorna Grameena Rozgar Yojana (SGRY)

Converging Jawahar Gram Samrudhi Yojana (JGSY) and Employment Assurance Scheme (EAS) a new scheme was launched called "*Sampoorna Grameena Rozgar Yojana*" during 2001-02. This scheme was launched with the intention of providing additional wage employment and food security in rural areas and to create durable community based and economic assets for infrastructure development in rural areas. The SGRY has two components viz.,

- 50 per cent by cash with 75% and 25% shared by the central and the state governments respectively.
- 50 per cent in terms of food grains from the Government of India (GOI), free of cost.

In the first stream, out of the 50 per cent allocation, 20 per cent was sanctioned to *zilla panchayats* and 30 per cent allocated to *grama panchayats*. Twenty two point five per cent of the allocation was separately earmarked for taking up works benefiting schedule caste and schedule tribes under first stream. Fifty per cent of the funds were allocated for the development of schedule castes and schedule tribe's colonies. The programme was implemented in all the districts. During 2001-02, JGSY and EAS continued separately along with SGRY till March 2002.

With the introduction of SGRY, the total allocation under EAS rose from 6947.48 lakhs to 10,020.93 lakhs and one lakh tons of free food grains. The allocation under JGSY rose from 8835.94 lakhs to Rs.9989.55 lakhs and one lakh tons of free food grains. With the advent of this scheme in the state, the state has benefited Rs.4227.06 lakhs in cash and 2.00 lakhs tons of food grains in kind. Out of this, the state Government provided Rs.1068.76 lakhs as its share of 25 per cent cash and Rs.900.00 lakhs towards transportation of food grains. The financial and physical performance of Sampoorna Grameena Rozgar Yojana (SGRY) both under I stream and II stream are shown in Figs. 3 and 4.

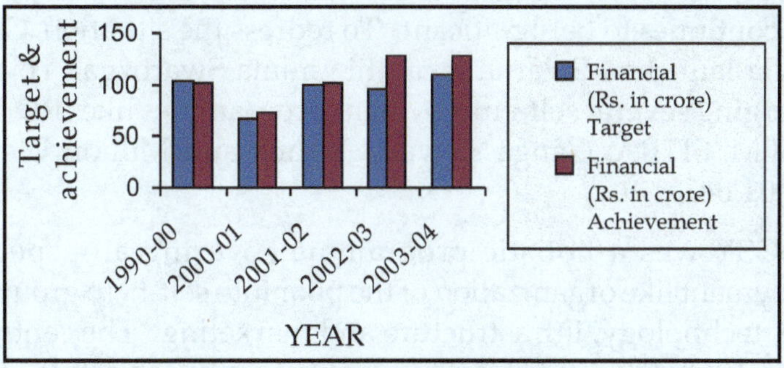

From ANONa (1999-2004)

Fig. 3. Financial Progress Under SGRY - I Stream

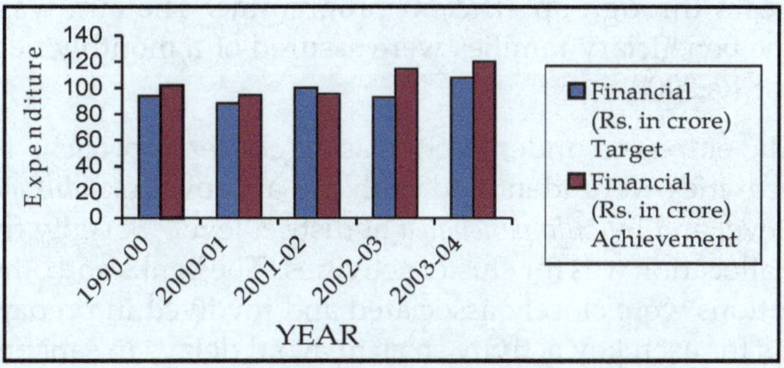

From ANONa (1999-2004)

Fig. 4. Financial Progress Under SGRY - II Stream

The first stream of SGRY was nothing but employment assurance scheme and the second stream was nothing but Jawahar Gram Samrudhi Yojana. The Sampoorna Grameena Rozgar Yojana (SGRY) under both the streams was successful in achieving more than the targets except food grain distribution in the II stream. In terms of physical man days created the achievement was overwhelming. To that extent poverty has been reduced in Karnataka.

2. SELF-EMPLOYMENT PROGRAMMES

The self-employment programmes aim to alleviate rural poverty by creating self-employment opportunities for the rural poor. A lot of stress was also laid on the creation of self-employment opportunities for women through these programmes.

2.1 Swarna Jayanthi Gram Swarozgar Yojana (SGSY)

Despite efforts made over the past few decades, rural poverty in India continues to be significant. To redress the situation, Government of India launched Swarna Jayanthi Grama Swarozgar Yojana (SGSY) by merging several self-employment programmes like IRDP, DWCRA, TRYSEM, SITRA, Ganga Kalyana Yojana and Million Wells Scheme from 01.04.1999.

SGSY was a holistic programme covering all aspects of self-employment like organization of the poor into self-help groups, training, credit, technology, infrastructure and marketing. The centre and state fund SGSY in the ratio of 75:25. Quality was the objective under SGSY. It aimed to bring poor families above the poverty line in three years and was targeted to cover 30 per cent of the poor in each block within five years through an efficient programme. The aim was to ensure that the beneficiary families were assured of a monthly net income of at least Rs.2000.

The emphasis under SGSY was on cluster approach. Four to five key activities were identified with the approval of *taluk panchayat* at *taluk* level and by *zilla panchayat* at district level. Seventy five per cent of the allocation was for cluster activities. The banks and other financial institutions were closely associated and involved in preparing project reports for each key activity so as to avoid delays in sanctioning loans and to ensure adequate financing.

SGSY sought to seek to promote multiple credit rather than one time credit injection. Training was imparted to the selected Swarozgaries about the programme and also in marketing strategies through well-designed training courses before providing assistance. Subsidy was uniform at 30 per cent of the project cost, a maximum of Rs.7500 for SC/STs, 50 per cent of the project cost subject to maximum of Rs.10,000. For groups Swarozagaries, the subsidy was at least 50 per cent of the project cost with a ceiling of Rs.1.25 lakhs. There was no monetary limit for irrigation projects. At least 50 per cent of the Swarozgaries were SC/STs, 40 per cent were women and 30 per cent was set part for the disabled.

Fifteen per cent of the funds under SGSY was set apart at the national level projects having a far reaching significance and which also acted as indicators of possible alternative strategies to be taken up in conjunction with other departments of Government or international organizations. This included initiatives to be taken in individual districts or across districts.

The programme was implemented soon after the project reports were prepared on the basis of the guidance received from the technical consultants and economists of the universities. SHGs were involved effectively to become self employed in the key activities identified viz: Dairy, sheep rearing, readymade garments, agro-food processing, flower cultivation, *agarabathi*, sericulture, carpentry, handicrafts, leather products, bamboo products, weaving, brick manufacturing, etc.

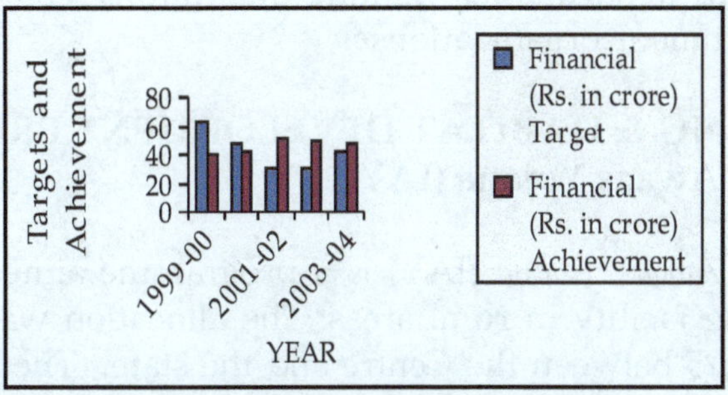

From ANONa (1999-2004)

Fig. 5. Financial Progress Under SGSY

The leading institutions like University of Agricultural Sciences, RUDSETIs, KEONICS and ATDC were also strengthened under infrastructure and utilized for training the Swarozgaries. For mass upliftment of poor, for each district one special project was proposed and five proposals worth of Rs.40 crores were sent to Government of India for sanction.

The financial and physical progress during 1999-2004 is given in Fig.5. It shows that in the first two years targets were not achieved. But during the later 3 years the achievements surpassed the targets. Similarly the physical progress of the scheme under the self-help groups formed and Swarozgaries assistance exceeded the targets in reducing un-employment through self-employment programme. SGSY programme was well received in Karnataka.

2.2 Self-Employment Programme–Swavalambana (SEP)

In addition to SGSY, the *Swavalambana Programme* was launched in rural areas during the year 1996-97 for providing self-employment opportunities to the educated young. The objective was to provide definite syllabus oriented, skill development training through reputed institutions, followed by financial assistance from banking institutions for setting up self-employment units. Training under *Swavalambana* covered both skill development and managerial aspects of the activity concerned. The scheme also provided for personality development to cover aspects like self-assessment, positive thinking, awareness on healthy living habits, developing confidence, improving intellectual ability, awareness of social and moral issues and responsibility towards community and society. *Swavalambana* involved cooperation of several Government departments, training institutions, credit agencies and non-Government organizations

3. HOUSING & HABITAT DEVELOPMENT PROGRAMMES

3.1 Indira Awaas Yojana (IAY)

Indira Awaas Yojana (IAY) is a programme aimed at improving the housing facility in rural areas. The allocation was shared in the ratio of 75:25 between the Centre and the state. The *taluk panchayats* implement the programme under the supervision of the *zilla panchayat*. The beneficiaries under IAY were selected by the *Grama Sabhas*. The responsibility of the programme was given to the *grama panchayats* from January 2004. Due to this the beneficiaries could get the benefit at the gram *panchayat* level itself instead of going to *taluk* headquarters.

The state government fixed a uniform unit cost of Rs.20,000 per house in the state on par with the unit cost fixed for the state schemes of *Ashraya* and *Ambedkar*. The houses had to be constructed by the beneficiaries themselves. The houses were allotted to either the women or jointly in the names of both wife and husband. The Government of India introduced a new component for up gradation of houses under Indira Awaas Yojana from 01.04.1999. Eighty per cent of the allocation of Indira Awaas Yojana was used for construction of new houses, while 20 per cent was used for the upgradation of houses.

The financial and physical progress of construction of houses and upgradation of houses during 1999-2004 is given in Figures 6 and 7.

The figures clearly indicate that the targets were not achieved during 1999-2004. Achievements were very less compared to the targets for each year. This is an indication that how far the people who are below the poverty line are provided shelter.

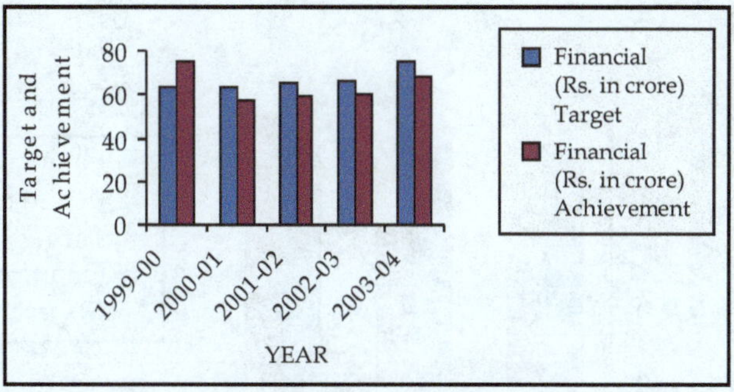

From ANONa (1999-2004)

Fig. 6. Financial Progress under IAY - Construction of Houses

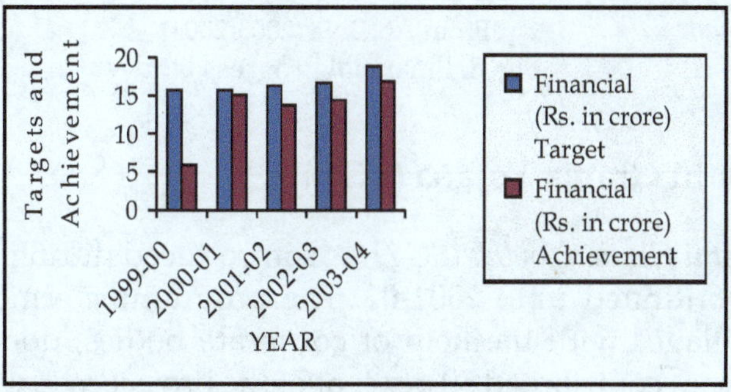

From ANONa (1999-2004)

Fig. 7. Financial Progress under IAY -Upgradation

3.2 Pradhanamantri Gramodaya Yojana–Grameena Awaas Yojana

Grameena Awaas Yojana was one of the five components of Pradhana Mantri Gramodaya Yojana along with primary health, primary education, rural drinking water and nutrition. This scheme was launched during 2000-01 in Karnataka state for providing shelter to the poor in the rural areas. The object of the scheme was to reduce the shortage of houses especially for families Below poverty line (BPL) in rural areas and also to facilitate a healthy development of the habitats in these areas. The progress made with regard to construction of village houses under PMGY is shown in figure 8. It indicates that, both financial and physical achievements except for the year 2003-04 were lagging behind the targets. The achievement was not even 30 per cent in the first year and was around 75 per cent in the second year; in third year the achievement exceeded the targets.

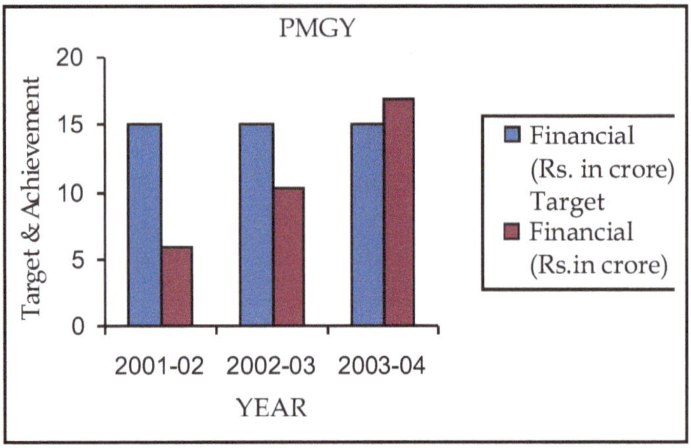

From ANONa (2003-2004)

Fig. 8. Financial Progress Under

3.3 Swasthi Grama Yojana (SGY)

Swasthi Grama Yojana (SGY) was introduced during the year 1997-98 and continued upto 2001-02. The programme aimed to develop model villages with the help of corporate bodies/donors. The state Government contributed 50 per cent of the project cost, while the donor was expected to contribute 40 per cent of the project cost. The remaining 10 per cent was contributed by the community. There was no separate allocation for the Swasthi programme. The *zilla panchayats* were expected to pool funds from various schemes for activities identified by the community. Thirty villages of 16 districts were identified as 'Swasthi Gramas'.

4. WATERSHED DEVELOPMENT PROGRAMMES

The Watershed Development Programme like drought prone area programme (DPAP), Desert development programme (DDP) and Integrated Watershed Development Programme (IWDP) was implemented by the *zilla panchayats* through watershed associations, the project-implementing agency, which could be a Government department or NGO. Each micro-watershed covered about 500 hectares. The PRI formed a watershed development team that interacted with the watershed association and provided technical assistance to the watershed association in the planning and implementation of the watershed programme. The residents of the area covered by the watersheds were also organized into self-help groups and user groups.

This approach, which was initiated in 1995, was in marked contrast to the earlier approach, wherein the line departments without the involvement of the community, implemented the watershed development programmes. Soil erosion, de-forestation and land degradation reduce the productivity of the soil and have an impact on the rainfall pattern. The ground water table also comes down leading to shortage of drinking water. These problems concern and affect the community. Therefore involvement of the community in planning and implementation of the watershed development programme was the key to the success of the programme. The process of creation of awareness among the community took some time and therefore the performance in watershed development programmes was affected to some extent during the initial years. The Government of India issued revised watershed guidelines as given below with effect from 1.1.2000:

1. The cost norm was enhanced from Rs.4000 to Rs.6000 per hectare.
2. The implementation of watershed duration was extended to five years.
3. Criteria for selection of watersheds primarily remained predominantly non-forest lands. The forest lands if happened to be a part of watershed, was also treated simultaneously.
4. Funding pattern was changed.
5. Monitoring and Review of programme was stressed.

The guidelines for watershed development were adopted with effect from 1995 and later on Government of India revised the guidelines in August 2001 to further simplify procedures and involve the *Panchayat Raj* Institutions (PRIs) more meaningfully in planning, implementation and management of economic development activities in rural areas. The new guidelines called "*Guidelines for Hariyali*" were issued and effected from 1.4.2003. Projects sanctioned prior to this continued to be implemented as per the guidelines of 2001.

4.1 Drought Prone Areas Programme (DPAP)

This programme concentrated on non-arable lands. It was implemented in 81 taluks of 15 districts. The allocation was shared in the ratio of 75: 25 between Central and State governments.

The progress made under this programme is shown in Figure 9. The progress under drought prone areas programme for the five year period indicated that from 2002-03, the achievement was more than the target whereas during the previous three years, the targets remained unfulfilled.

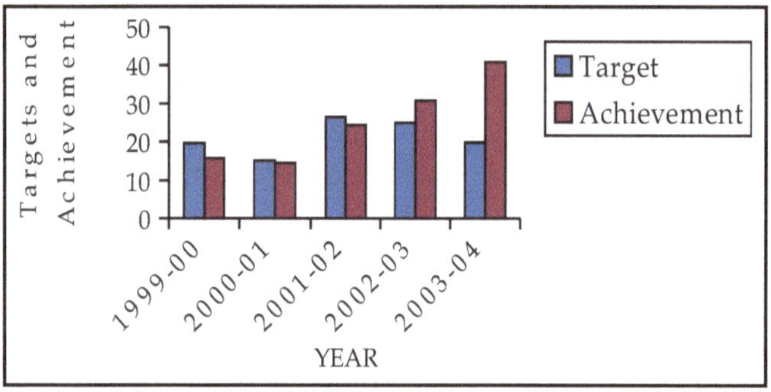

From ANONb (1999-2004)
Fig. 9. Financial Progress Under DPAP

4.2 Desert Development Programme (DDP)

This programme focused on afforestation to arrest the growth of hot and cold deserts and was implemented in 22 *taluks* of 6 districts. The allocation was shared in the ratio of 75: 25 between Central and State governments. The progress achieved under this programme is shown in Figure 10. The achievements were below target during first and second year, met the targets in the third year of implementation and surpassed targets during 2002-2003 and 2003-2004 (Fourth and Fifth Year).

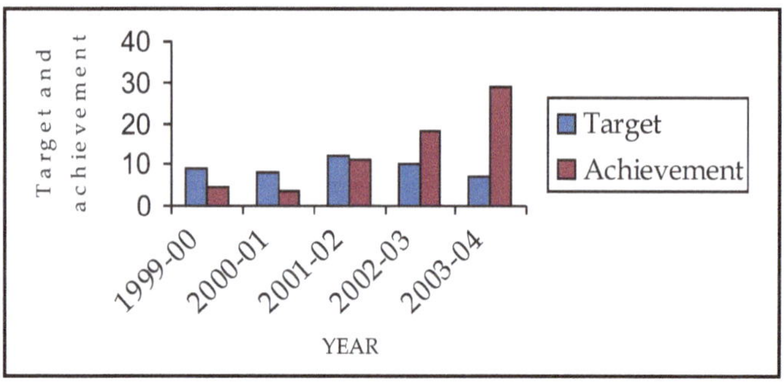

From ANONb (1999-2004)
Fig. 10. Financial Progress Under DDP

5. RURAL WATER SUPPLY AND SANITATION

Availability of safe drinking water in sufficient quantity to 56,682 rural habitations of the state was given utmost importance in order to improve the living standards of the rural masses. Potable drinking water was supplied to rural areas through three types of programmes namely.

1. Bore wells with hand pump scheme.
2. Mini water supply scheme.
3. Piped water supply scheme.

The rural water supply programmes were implemented as per the transparency principles of the state through *zilla panchayat* Engineering Divisions. Priority was given to the identified problematic villages. The type of water supply scheme to be taken up was decided on the basis of the total population of the village/habitation. Village/ habitation with a population of 1000 and above was provided with piped water supply scheme. Village/habitation with a population between 500 to 1000 was provided with mini water supply scheme. The other village habitations with less than 500 population were provided with bore wells fitted with hand pumps. The above criteria was relaxed in deserving cases where villages/habitations occurred in hard-core areas and having water quality problems like high fluoride contents, etc.

5.1 Bore Wells With Hand Pumps Schemes

Bore wells fitted with hand pumps are the major source of potable drinking water in rural areas. Since the inception of the programme to the end of March 2004, 1,87,818 bore wells with hand pumps were provided in the state. During 2003-04, it was planned to install 6,364 bore wells with hand pumps at a cost of Rs.25.45 crore under all sectors. During 2003-04, 4,111 bore wells have been installed at a cost of Rs.22.89 crore. An amount of Rs.600/- per bore well was provided for their annual maintenance. *Taluk panchayats* took care of these bore wells.The financial and physical progress achieved. with regard to bore wells with hand pumps during the period of implementation is shown in figure 11. It shows that both financial and physical targets were not achieved and to that extent people were deprived of getting safe drinking water.

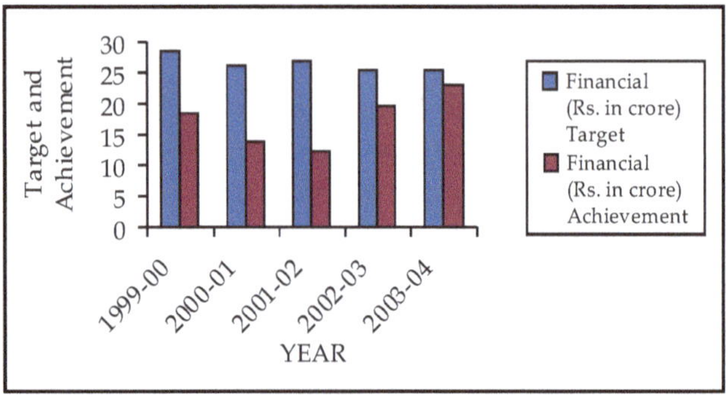

From ANONb (1999-2004)

Fig. 11. Financial Progress Under Bore Well and Hand Pump Scheme

5.2 Mini Water Supply Scheme

In this scheme, water is pumped to a small tank (cistern) fitted with 3-4 taps from where water can be collected by household. Since the inception of this programme to the end of March 2004, 21235 mini water supply schemes have been completed and commissioned.

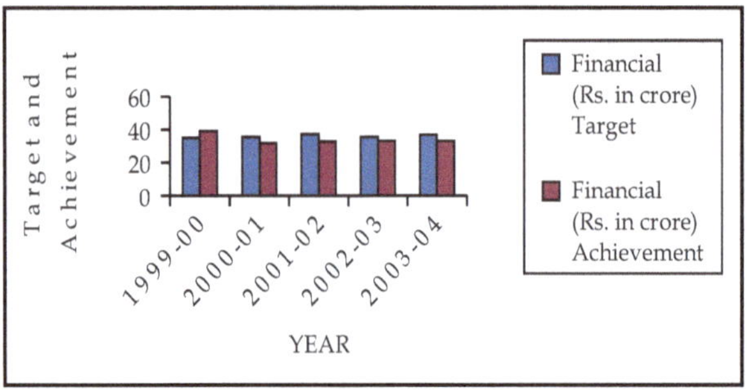

From ANONb (1999-2004)

Fig. 12. Financial Progress Under Mini Water Supply Scheme

During 2003-04 it was programmed to take up 1245 mini water supply schemes both under state and central sectors at a cost of Rs.36.82 crore. The achievement till the end of March 2004 was 1256 schemes at a cost of Rs.33.35 crore. *Gram panchayats* are maintaining these schemes. For this, an amount of Rs.3500/- per annum was provided to each mini water supply scheme.

The progress of Mini water supply scheme in terms of financial and physical achievements for the last five years is shown in Figure 12. The financial targets were not achieved except during the year 1999-2000.

5.3 Piped Water Supply Scheme

Under this programme, since the inception to the end of March 2004, 16375 schemes were completed and commissioned both under state and central sectors. During 2003-04 it was planned to commission 725 piped water supply schemes at a cost of Rs.58.02 crore. Of this, 635 piped water supply schemes have been completed and commissioned by end of March 2004 at a cost of Rs.48.46 crore. *Gram Panchayats* are maintaining piped water supply schemes, with an amount of Rs.8000/- per piped water supply scheme per annum.

The financial and physical progress under the piped water supply scheme in Karnataka is shown in figure 13. Physical progress under the scheme is not satisfactory except in the year 1999-2000 when the number of piped water supplied scheme in Karnataka exceeded the target.

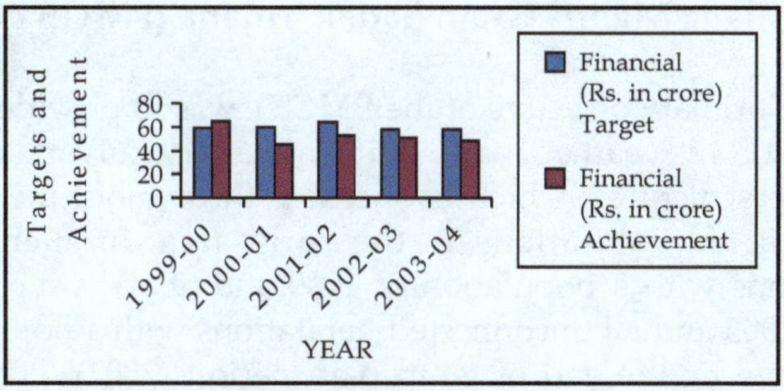

From ANONb (1999-2004)

Fig. 13. Financial Progress Under Piped Water Supply Scheme

5.4 Prime Minister Gramodaya Yojana–Rural Water Supply

Government of India introduced a new initiative under Prime Minister Gramodaya Yojana for the sustainability of drinking water sources in rural areas from 1999-2000 onwards. The funds were utilized for construction of schemes such as water conservation, rain harvesting structures and water recharge structures viz., construction of check dams, desiltation of tanks, percolation tanks, infiltration well, *Gokatte*, etc. The action plans prepared by *zilla panchayats* were approved at Government level and schemes implemented in all the 24 districts except Mysore, Bellary and Dakshina Kannada, which were covered under sector reforms project. The progress achieved under this scheme during the planned five years with regard to financial and physical progress is shown in figure 14. The financial and physical progress clearly indicates that the physical achievement was not satisfactory except during 2003-2004.

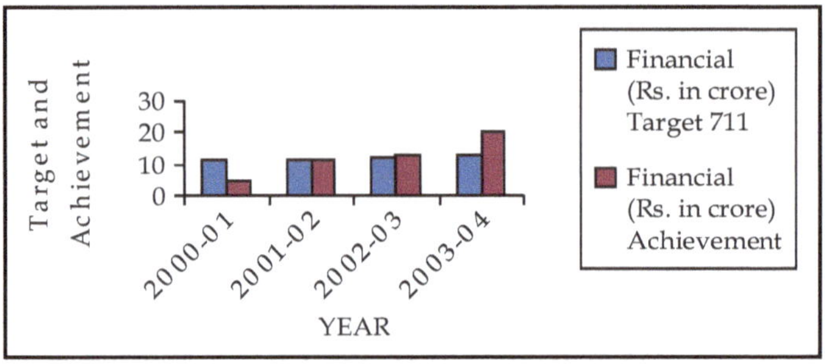

From ANONb (2000-2004)

Fig. 14. Financial Progress Under PMGY - Rural Water Supply

5.5 Pradhana Mantri Gram Sadak Yojana (PMGSY)

The primary objective of the PMGSY was to provide connectivity by way of all weather roads (with necessary culverts and crossing drainage structures, which is operable throughout the year), to the unconnected habitations in the rural area, in such a way that habitations with a population of 1000 and above are covered at the end of 2003 and all unconnected habitations with a population of 500 and above by the end of tenth plan period (2007). The Karnataka land army corporation and the *zilla panchayat* engineering divisions have been designated as the executing agencies in the state. The upgrading work of 2,663 kms of all weather roads is under progress. Up to the end of March 2004, Rs.190.32 crore of expenditure has been incurred and 477 works were completed, 2014 kms of road length was upgraded to all weather roads and rest of the roadwork is in the final stages of completion.

5.6 Nirmala Grama Yojana (NGY)

Nirmala Grama Yojana (NGY) was implemented in the state with effect from 2nd October 1995. Keeping in view the hygiene and improvement of living conditions of rural mass, under this programme more than 9 lakh household toilets were constructed. A subsidy of Rs.2000/- for BPL and Rs.1200/- for ABPL category family was given for the construction of each toilet. During 2003-04 an amount of

Rs.11.27 crore was provided out of which Rs.7.26 crore was spent for the construction of household latrines and the balance amount was spent towards rural schools and sanitation programme.

Figure 15 shows the financial and physical progress made under NGY in Karnatkaa. It indicates that progress was behind schedule during the entire duration of implementation (1999-2004)

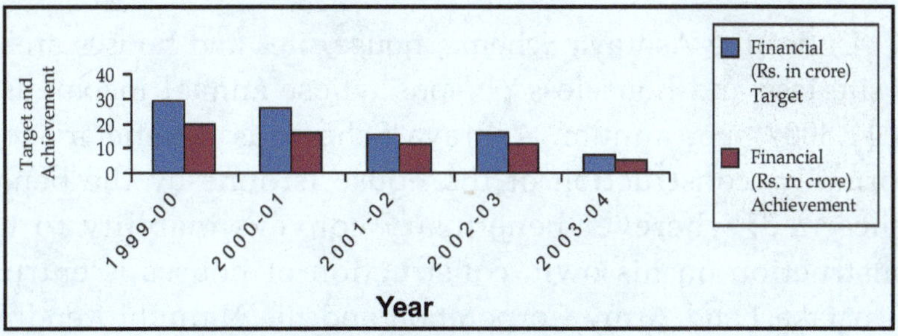

From ANONb (1999-2004)

Fig. 15. Financial Progress Under NGY

In addition to the above rural development programmes, there are rural energy programmes like National Project on Biogas Development, Integrated Rural Energy Programme etc.

Further, there are some externally aided and centrally sponsored rural water supply and sanitation projects. They are Netherlands Assisted Integrated Rural Water Supply and Sanitation Project, World Bank Assisted, Karnataka Integrated Rural Water Supply and Environmental Sanitation Project, World Bank assisted Second Karnataka Rural Water Supply and Sanitation Project, 'JAL NIRMAL' project (Danida assisted rural drinking water supply and sanitation project), 'JALNIDHI' (Danida assisted rural drinking water supply and sanitation project -phase-II), Centrally sponsored Swajaladhara Project, sector reforms, community based rural water supply programme and centrally sponsored total sanitation campaign.

6. KARNATAKA HOUSING SCHEME

The Government of Karnataka has traditionally given high priority to public housing. It is one of the few states, which has its own housing programmes for the economically weaker sections (EWS) both in rural and urban areas apart from the centrally sponsored housing schemes for the economically weaker sections. In addition, the Karnataka

Housing Board caters to housing needs of low income, middle income and high-income groups. Established by the Government of Karnataka on 25.1.2000, it caters to the housing needs of the economically and socially weaker sections of society.

6.1 Ashraya Scheme

Under the Ashraya Scheme, house sites and houses are allotted to site less and houseless persons whose annual income is below Rs.11,800/- per annum. Ashraya Scheme is beneficiary-oriented. Normally, construction of the house is done by the beneficiary concerned; wherever beneficiary conveys inability to take up construction on his own, construction of houses is entrusted to Karnataka Land Army Corporation and the Nirmithi Kendras. The use of contractors is banned. The implementation of the scheme has been entrusted to Rajiv Gandhi Rural Housing Corporation Limited, which operates through the Chief Executive Officers of *zilla panchayats* at district level in respect of rural houses and through the Deputy Commissioners of districts in respect of urban houses.

6.2 Distribution of House Sites (Ashraya Sites)

During 2003-04 under Rural sites programme and the distribution of house sites under Ashraya scheme, a budgetary provision of Rs.4.59 crores was made for both rural and urban sites schemes under house sites programme and a total Rs.14.50 crores was released to Rajiv Gandhi Rural Housing Corporation Limited.

6.3 Rural Ashraya

During 2003-04, 108747 houses were constructed under this scheme. This also included backlog houses of previous year constructed during this year. For this purpose a total of Rs.85.00 crores was released to Rajiv Gandhi Rural Housing Corporation Limited as subsidy, as against a budgetary provision of Rs.95.00 crores from Housing Department. Rupees Twenty crores from Social Welfare Department was also released during this year.

6.4 Rural Ambedkar

During 2003-04, 16274 houses were constructed under this scheme. This included backlog houses constructed also. For this purpose a total of Rs.18.83 crores was released to Rajiv Gandhi Rural housing Corporation Limited as subsidy.

The state housing programmes implemented during the last 4 years have had a significant and substantial socio-economic impact on the community in terms of better shelter, leading to improved esteem, literacy, health, hygiene and sanitation, environment and crime reduction. Well laid out habitations with access to roads and drains, drinking water, electricity and a green environment have improved living conditions. The progress made under this programme is satisfactory only in Bangalore urban. The progress in Chamarajanagar, Chikamagalore, Chitradurga, Kodagu, Shimoga and Udupi was more than the target. Whereas, in the remaining districts of Karnataka, the progress was not satisfactory.

Self-help Groups: One of the primary goals of the government of Karnataka is to eradicate poverty and empower women. The eradication of poverty was attempted via several poverty alleviation schemes. These schemes generally targeted the poorest beneficiaries and provided credit for them. The latest approach in eradicating poverty is via self-help groups for women. These women groups are encouraged to save money. The Government steps in with revolving fund as well as bank credit; this method is found to be the most effective in delivering rural credit as well as eradicating poverty.

DISCUSSION

It is becoming more and more evident that the majority of the poor in developed and developing world are women. Poverty among rural women is growing faster than among rural men. Over the past 20 years, the number of women in absolute poverty rose by 50% as against some 30% among rural men. The alarming evidence concerning the underlying trends for this process strongly indicates that the gender composition of the poor is veering towards a greater share of women. The lack of progress in the past twenty years in the eradication of poverty and growing proportion of women among the poor is the single most important threat to the progress of development and its sustainability. As long as three-quarters of the world population continue to suffer from acute deprivation, as long as profound imbalances in global consumption continue to persist, and more important, as long as the spread of poverty, particularly among women, continues unchecked, there can be no development. The history of the development process shows again that the economic status of women is the key variable in the solution to poverty crisis. It is time for the full recognition of the fact that women are part of the solution to poverty and to the stagnating development, not part of the problem.

With decentralization of governance, decisions regarding implementation and budgetary allotment are made at district, *taluk* and village levels. There is lot of transparency in project implementation. This coupled with reservation for women and SCs & STs in *zilla* and *gram panchyats*, is intended to lead to inculcation of gender sensitivity in development programmes. With self-help groups liberating women from the clutches of money lenders, farm women are bolder, less worried and are well aware of their potential in changing their life towards better standards and more comfort. The latest arm to fight for justice is the passing of Right to Information Act. Endowed with economic liberty, reservation, awareness about rural development programmes and now with a right to seek any information, rural women are on their way to empowerment. What they need is good leadership initially and training to become leaders in their own right. It is intended that the light thrown on the various governemnt schemes aiming at rural development in this paper will help extension specialists and agricultural scientists in making farm women more concious of their rights and enhance their ability to monitor the progress of these programmers to improve their lives.

References

ANONa. 2003-04. Annual Report, Department of Rural Employment and Panchyat Raj, Government of Karnataka, Bangalore.

ANONb. 2003-04. Annual Report. Housing Department, Government of Karnataka, Bangalore.

15

Role of Women in Dairy Farming

DHANANJAYA, B., R. VINAY KUMAR
AND C. CHANDRASHEKHAR

ABSTRACT

Rural Indian women are extensively involved in agricultural activities and play a major role in agrarian scenario. Majority of the rural farm women are actively involved in various operations related to mixed farming system existing in India. Women and men participate in agricultural and animal husbandry practices as a family unit on the farms. Men perform major share of work quantitatively on agriculture side, where as on the animal husbandry side most of the management operations are being carried out by women in addition to their household activities. Dairy production is less influenced by rain failure compared to crops and is invaluable in sustaining family income during drought. Dairy development is now considered as an important tool for rural development or poverty alleviation. In fact, dairy farming occupies a special place in rural society with its linkage with religion and culture apart from providing sustained income. Most of the cattle and buffaloes are commonly kept in every farm household primarily for socio-economic reasons and more importantly for gaining drought power and as a secondary source of nutrition subsistence in the form milk and milk products for the whole family. In this context, a study was conducted in Karkala *taluk* of Udupi district with the major objective of studying the form and extent of participation of farm women in various operations of animal husbandry. The attempt made in this regard has revealed that rural women are fully involved in majority of the operations either as manual workers or as supervisors. Hence, it is suggested that for massive empowerment of rural women, animal husbandry programmes need to be launched as they are inherently profitable.

16

Entrepreneurial Development of Self Help Groups in Agriculture Role for Women Scientists

DHAKSHAYANI, C. & SULTAN AHMED ISMAIL

ABSTRACT

Entrepreneurial training was offered to village women in the field of agriculture especially for women members of Self Help Groups. These women were trained in the production, utilization and marketing of Bio dung compost and vermicompost from agricultural wastes. They were also trained in the production, utilization and marketing of vermiwash (a liquid foliar spray), *Panchagavya,* biopesticides and herbal pest repellents using locally available plant materials. An attempt was also made to educate the people on the importance of food and nutrition through diet counselling and the importance of nutrient (kitchen) gardens. They were trained about Circular (Ring) Gardens and Double Digging gardens. The women were given these trainings as part of converting them from dependence on chemicals in agriculture to organic farming. There was overwhelming response, and the women now use these agricultural products in their own fields. The role of women scientists in this programme shall be elucidatd during the presentation of this paper.

When a man is educated, he alone is benefited, but when a woman is educated her family and the society are benefited. Proving these, the women in the Poorivakkam village are attempting to convert the whole village into organic farming. Their organic flowers have carved a niche for themselves in the Chennai flower Market.

17

Bottlenecks to Become A Breadwinner

SAILAJA, A

ABSTRACT

Formulation of any action plan for women needs understanding of their access to and control over the resources. To have deep insight into this aspect, a study was undertaken with 240 small and marginal farm women randomly selected from eight villages of four districts in Andhra Pradesh in consultation with officials of agriculture department. Using an interview schedule, impediments for development of farm women (social, economic, psychological, extension and technical) were collected and prioritized through matrix ranking.

To overcome social and economic constraints, gender mainstreaming and budgeting are now recognized as the most important strategies for women's empowerment at all stages of any developmental programmes. Such an approach necessarily aims at achievement of gender equality. Organizing capacity building programmes in managerial, entrepreneurial, negotiating and leadership skills improves risk taking ability (psychological) and enhances active participation of farm women in developmental activities.

Extension constraints can be overcome by deployment of more women extension trainers, introducing drudgery and time relieving technologies which also enable her to participate in extension activities. These have to be conducted in village itself. Technical barriers in development can be crossed through provision of crèche facilities at training centres, following holistic and integrated approach and giving training in self employment ventures (bio input preparations) with assured marketing. This provides gainful employment to farm women besides increasing cost effectiveness of production as well as environmental soundness. This study enables us in strategy formulation for farm women's empowerment through promotion of agri-business sectors in rural areas and thus achieve 4% growth rate in agriculture.

18

Income Generation through Vermicomposting Among Rural Farm Women

BIRADAR, A.P., D.P.BIRADAR & A.K. GUGGARI

ABSTRACT

Major occupation of women in Bijapur area is agriculture with an average land holding of 1-2 ha. Most of them are unaware of advantages of vermicompost. In recent times, dependence on agriculture is becoming very difficult for them. Majority of the women population in these areas are below poverty line (BPL), because of increased cost of production, and fluctuations in market prices of major crops grown by them. Erratic and ill distributed rain fall pattern makes life hard for them. The ill effects of chemical fertilizers have lead to interest in organic farming through vermiculturing using farm and urban wastes in order to produce risk free crop at reduced cost of cultivation.

Considering these problems, a survey was made in Bijapur district covering Kumathe, Nidoni, Kadlewad, Baganagar, Ukkali and Hattarakihal villages to know the economic status of rural women and encourage them to involve in income generating activities to improve their economic status by practicing an economically feasible activity i.e. vermicomposting during 2005-06.

Results of the study revealed that, practice of vermicompost production and selling/or for own use increased the family income substantially. The technology demonstrated that practitioners of vermicompost became self sufficient to meet organic manure requirement of their fields. They also got additional income of Rs. 18,600/annum in one *gunta* of vermicompost production area.

19

Entrepreneurship Development A Way to Rural Women's Empowerment

KALPANA, B., K.V. JAMUNA & B. KRISHNAMURTHY

ABSTRACT

Empowerment strategy is today emerging as a unique Indian response to the challenge of equality and development. As per the population census, women constitute more than 50 percent of total population, but rural farm women are the most deprived in today's world. Farm women are poorest among poor. Even though they work the longest hours on farm and have achieved more with lesser resources, they are not recognized and appreciated for their contribution. In India 78 percent of all economically active women are engaged in agriculture. Almost 50 percent of rural female workers are agricultural labourers and 37 percent are cultivators. The rural poor also demand development through creating income generating activities for women in rural areas. Women empowerment gained momentum after independence through different governmental programmes such as education, creating income, self help groups, subsidy programs to women etc. Women can derive income from subsidiary enterprises such as dairy, duckry, poultry, etc., and preparation of vermi-compost, vegetable cultivation, preparation of jam, jellies, *papads*, pickles, embroidery, tailoring, honey bee keeping, kitchen garden, value addition and food processing etc. But marketing poses a great problem for these products. Based on our observations and experience gained as extension specialists, suggestions are made for women empowerment in this study.

20

Empowerment of Women through Some Agrobased Activities

AFIFA S.KAMILI

ABSTRACT

Women emancipation, women liberation, women awareness, women upliftment, women empowerment, gender equity etc. are the buzz words our media cashes on. But the question remains: What has made us all to talk about her in 21st century? Unfortunate, but true that women, constituting one half of the total world population, have been treated as second class citizens and denied even the basic rights. Men have dominated her by neglecting her talent, capabilities and potential. Many measures were undertaken by the Government for the betterment of the women such as the creation of National Commission for Women, Parliamentary Committee on Empowerment of Women, Gender Budgeting besides creating many disciplines in various institutions/Universities for women studies and passing several pro women laws to protect her rights. But this is not enough. What is needed is the freedom from disability of any kind lying in the minds of various sections of the society. A woman need not be pitied but given the due she deserves.

Historians believe that it was women who first domesticated the wild food crop and initiated the art and science of farming. As per the World Economic Profile, they form 50% of world population, contribute 60% work force, making up to 30% of the official labour and contribute 50% in food production. At the home front, nearly 84% of all the economically active women in India are engaged in agriculture and allied activities. Even there are certain operations in agriculture in which women dominate as in production, post harvest management and agro processing.

The largest number of women in India live in rural areas and majority of them are engaged in farm activities. The rapid increase in population coupled with decrease in agricultural land has created an alarming situation. Therefore, it is strongly felt that women need some agri based entrepreneurship which can supplement their family income, make them economically independent and improve their status. Various agro based activities for women entrepreneurship have been identified. These include processing of spices, condiments, milk and milk products, meat, fish, chicken, extraction of seeds from fruits and vegetables, preparation of snack foods, mushroom culture, sericulture, frozen vegetables, fruit juice, nectar making, dehydrated onions, potato chips/wafers etc. It is believed that once we are able to instill confidence in the rural women entrepreneurs for making quality produce, the rural scenario will change beyond imagination. Let us hope that 50% of our population are given full support at all levels to realize their full potential to make the agri-based activities rewarding for her.

❑❑❑

SECTION-4

CAPACITY BUILDING

21

Capacity Building through Empowerment–Sthree Shakthi An Experience in Karnataka

NARAYANA GOWDA, K, K.JAGADEESWARA, L.RAMAKRISHNA RAO, A.K. CHAKRAVARTHY AND M. RANGANATH

ABSTRACT

The Stree Shakthi Groups (SHGs) were launched in Karnataka during 2000-01 aimed at improving the overall development of rural women in Karnataka. A review of the progress achieved by these illeterate or less educated, hither to continued to be economically dependent with no voice in decision making, is nothing short of a miracle. With an enrollment of one and a half million in 5000 villages, their savings and borrowings run into hundreds of crores and employment opportunities are provided to lakhs of farm women. Apart from agriculture related activities, some of them have also involved in some unconventional, income generating activites. Economic empowerment has emboldened these beneficiaries to not only fight against social evils but also win their battles. Overall, SHGs have transformed the life of rural women treatmendously in a positive direction.

INTRODUCTION & OBJECTIVES

A meaningful and a tangible programme was launched in Karnataka in 2000-01 for furthering social, economic and overall development of rural women named 'Stree Shakthi' with an idea of creating self-employment for income generation, social security and equity through capacity building empowerment.

Objectives of Stree Shakthi Programme

- To accelerate the process of economic development and create conducive atmosphere for social change of rural women.
- To form women self-help groups based on thrift and credit principles
- To increase the income levels of rural women by engaging them in income-generating activities and creating financial stability and thereby achieving eradication of poverty.
- To converge services and benefits of various Government departments and financial institutions.

Stree Shakthi Groups (SSGs) are formed through *Anganawadi* workers appointed at village level. Each self-help group consists of 15 to 20 like-minded women coming from the lower strata of society having similar social and financial background. With a sense of oneness they organize themselves voluntarily into a group for their social and economic development.

RESULTS & DISCUSSION

As on 2004, one lakh SSGs were formed enrolling 15.09 lakh women spread over roughly in 5,000 villages. The cumulative saving of SSGs amounted to Rs.254.39 crores and they have availed Rs.212.48 crores as bank loan. Out of one lakh SSGs, 56,992 SSGs have started economic activities providing either full time or part time employment to more than 3 lakh rural under privileged women. While 81 SSGs have saved an amount of more than one lakh rupees, 243 SSGs have saved Rs. 50,000 to one lakh from each group so far.

Irrespective of the different regions in the State majority of members of SSGs have taken up mainly agriculture related activities such as dairy, sheep/goat rearing, poultry, etc. The non-agricultural activities were tailoring, processing foods and petty business.

Apart from agricultural related activities, it is interesting to note that some of the non-traditional enterprises such as running canteen, cloth shop, saree sales, trading, bakery, manufacture of Rexene bags, etc., were the identified and preferred vocations by sizable number of groups.

In addition to creating opportunities for self-employment and income generation, SSGs have made greater strides and have created a movement in rural parts of the state. It's impact is manifested in many ways. Some of the women SSGs are protesting against the sale of arrack and liquor in their villages and have become successful in closing them down. SSGs are also fighting against social evils like child marriage, dowry, gender inequality and domestic violence. There are reports that these women SSGs are taking active part in community development. In brief, it can be said unhesitatingly that rural Karnataka is witnessing a quiet revolution and SSGs have succeeded in brining a major transformation in the lives of lakhs of rural women. Considering the experience of SSGs of Karnataka, in future more attention should be given to activities which would enable the women groups mainly by way of capacity building intellectually, socially and economically.

22

Women Training as A Challenge for Agriculture Development in India

ASHWINI, A., B.K. NARAYANA SWAMY
AND G.U.BHARATHAMMA

INTRODUCTION

Women constitute half of the humanity, contributing to two-thirds of world's work hours. She earns only one-third of the total income and owns less than one-tenth of the world's resources. This shows that the economic status of women is in a pathetic condition and this is more so in a country like India. Rural women in India share abundant responsibility and perform a wide spectrum of duties in running the family, maintaining the house, involving in activities like rearing, feeding livestock, attending to farm labour, and the likes. Still they suffer economically and are socially invisible. Training of rural women is important so as to increase their involvement in developmental process, enhance their skill and make them equal partners in national development. The major objectives of training of rural women should be to equip them with better skills and enhance their knowledge so as to prepare them to face new challenges due to technological developments. Education, social participation, extension participation, training, mass media use and capacity building has significant association with the level of empowerment, whereas age, family size, land holding and material possession had no association with the level of empowerment. However, training is concerned with increasing knowledge and skill in doing a particular job. We advocate training as an intensive learning activity for a group of selected rural women assisted by competent trainers to understand and practice the skill required in adoption of new technology at a place where appropriate facilities exists and at a time and duration considered suitable by the farm women for their empowerment.

Technology Development For Farm Women: There are four essential components of Women Resource Development (WRD) that is production, equity, sustainability and empowerment. Women in India must participate fully in decision-making process to shape their lives. The system dealing with production technologies for rural women in the country failed to meet the fundamental needs of rural women because of its inappropriateness and low functioning efficiency. Several technologies developed through research were not tested under real farming conditions in the country. Further, rural women and extension workers were not involved in technology generation process resulting in inappropriate technologies. Thus, there is an urgent need for a model to develop appropriate technologies for women to participate effectively in agriculture development. This technology development process should cover all the requirements with proper decisions at all stages of activities. The model should consider all the minute details needed for technology development. Further, the technology should have certain attributes for its wider adoption. Narayana's Lotus Model (Fig.1) advocates that an appropriate technology for rural women should have four major attributes with seven sub items under each head for agriculture development like *Relative advantage* (Market feasibility, Market stability, Cost, Net profit, Profit consistency, Utility potential and Time saving), *Compatibility* (Cultural, Social, Physical, Psychological, Situational, Relational and Anticipated), *Practicability* (Point of origin, Access to advice, Visibility, Mastery, Trail ability, Demonstrability and Communicability) and *Complexity* (Failure probability, Discomfort saving, Resource complexity, Reversibility, Work efficiency, Cognitive complexity and Application complexity). The sustainable, productive technologies for rural women should be based on agricultural research. Hence, agricultural universities and research stations should give due consideration for developing appropriate technologies to suit needs of rural women for agriculture development.

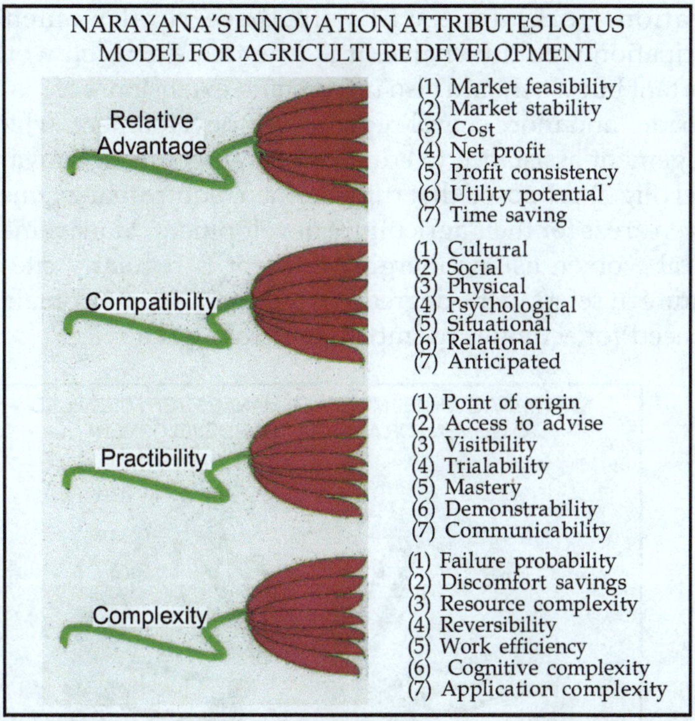

Fig. 1. Appropriate attributes of Modern Agricultural Technologies are shown on the lotus.

Providing Management Service to Rural Women: Our studies emphasized that four important kinds of services are needed by rural women for adoption of emerging agricultural technologies. These are (1) Engineering help in building basic physical infrastructure (2)Technical advice and knowledge bearing on planning and management (3) Economic and logistical help in securing credit and inputs (4) Social services including education. Under the Government programs most of the four essential services are being strengthened. In spite of these facilities the services are not reaching the rural women in the absence of integrated management services. Although the major emphasis has been on effective utilization of emerging agricultural technologies by rural women, the spread of adoption of technologies in the country has been extremely slow. The unfortunate experience of several countries is that in spite of providing various kinds of services to rural women at local level, these services cannot be made to function adequately and effectively without an integrating management service operating at village level. The importance of such a type of integrating services is not recognized fully so far. Assuming that extension education work alone will help these rural women for their participation would be naive. It is true that extension work is vitally important but we should also understand extension work is an element of broader and more complex process of development, which requires management assistance to educated rural women. Narayana's Wheel Model (Fig.2) proposed that rural women require management services in seven areas for their agriculture development. Management services to rural women as an emerging concept is regularly cited in recent literature. It refers to the degree to which rural women obtain assistance they need for.activities to implement technical advice.

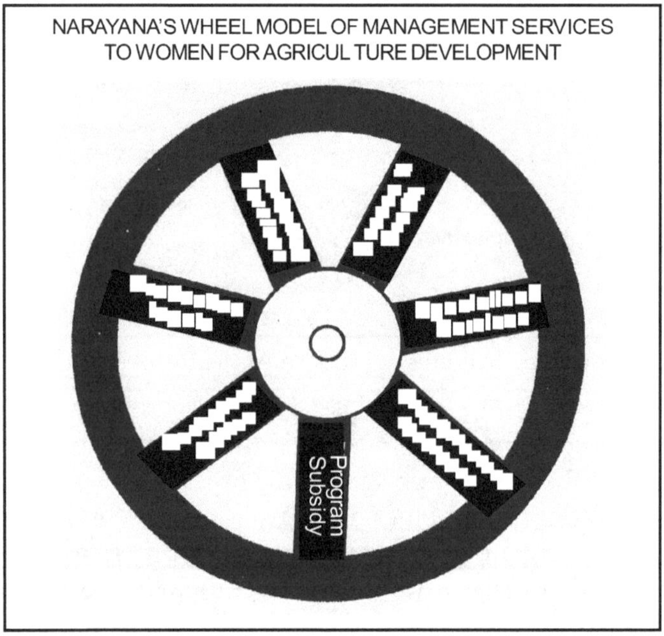

Fig. 2. Provided Management Services to Seven Orchestrated, Concerted, Comprehensible Operational Activities in the agricultural system are shown on the wheel

Skill Development Among Women through Training: Indian women get only 25% of the share in the earned income, while their male counterparts take-home three times that giving a clear indication of the disparity in their status and empowerment. Educational backwardness is the major reason for women lagging behind men in India. Hence there is need to develop skill among rural women for agriculture development of the country. Skill plays a very important role i.e. an acquired power of doing something comparatively better.

Different activities demand different skills the relative importance of which may vary with the job undertaken. Narayana's Wheel Model (Fig. 3) describes seven skills required by rural women for efficient agricultural development. These are: (1) Technical skill is the ability of the rural women to use any technique or method or equipment or product or process as a tool in the context of agriculture i.e. all farm inputs, practices, process of services to achieve target in agriculture. (2) Human Skill is the ability of rural women in motivating other people involved in agriculture with thorough understanding while working with them as a team. (3) Conceptual Skill is the ability of rural women in co-ordinating and integrating all the activities of farm interest with visionary outlook. (4) Managerial Skill is the ability of rural women in planning, organizing, directing, leading, reporting, budgeting and reviewing the work of other people involved in agriculture. (5) Design Skill is the ability of rural women in finding out a workable solution to problems. It requires deliberate efforts to develop solution. (6) Creative Skill is the ability of rural women in generating new ideas or in doing things already done in a new way in agriculture. (7) Communicative Skill is the ability of the rural women to adopt farm technologies at different levels using series of methods over a period of time.

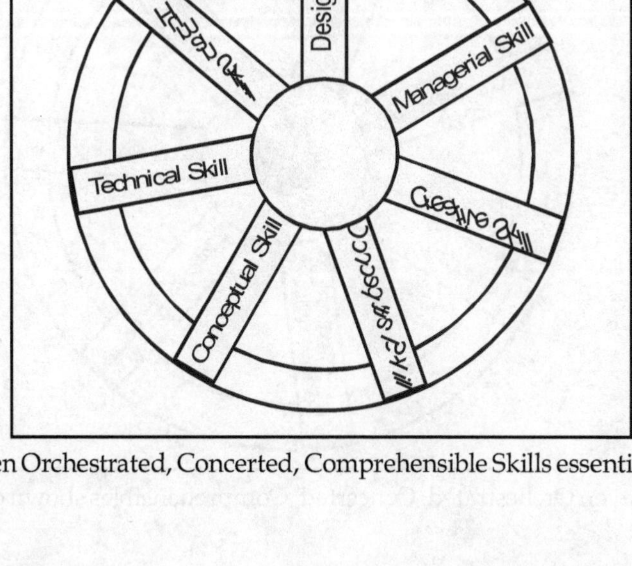

Fig. 3. Seven Orchestrated, Concerted, Comprehensible Skills essential shown on the Wheel

Capacity Building Among Women Through Training: There is a great need for capacity building among rural women for agricultural development in India. The capacity building models among rural women in future will have to focus on (1) Initiate (2) Establish (3) Build (4) Develop (5) Manage (6) Supply goods and (7) Specialize in their own strength. However, capacity building among rural women is too complex a phenomenon to be explained by a single factor. It is a combination of several factors. So far limited attention is given to promote capacity building among rural women. The research on capacity building is also scanty. The worldwide bibliography and research reports and studies on capacity building are very limited. Some of the characteristics that are based on research from different countries of the world are not cross culturally valid. Hence, the characteristics which distinguish capacity building may not be uniform for different strata of society. Narayana's wheel Model (Fig.4) explains combination of seven components for capacity building among rural women viz., Innovativeness, Decision making ability, Achievement motivation, Information seeking ability, Risk taking ability, Coordinating ability and Leadership ability.

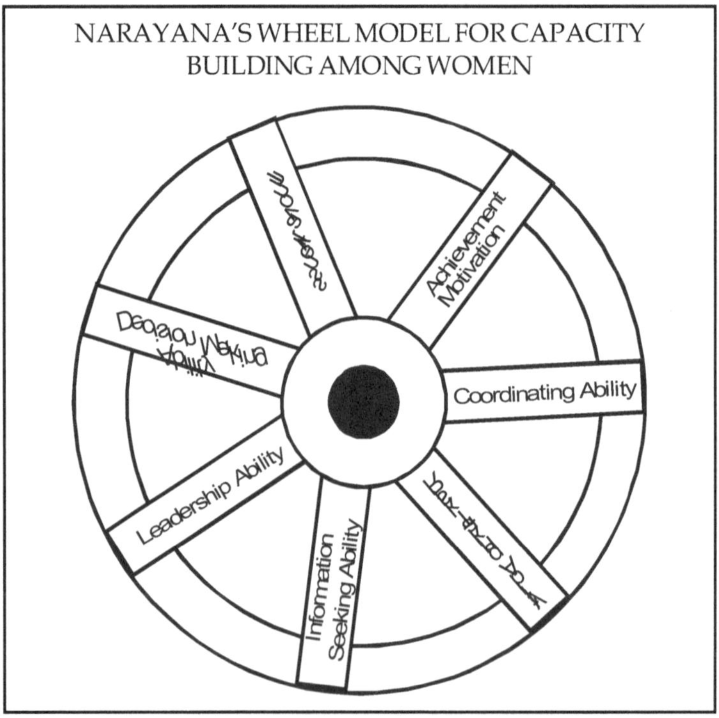

Fig. 4. Seven Orchestrated, Concerted, Comprehensibles shown on the Wheel

Impact Assesment of Training Women for Agriculture Developmen: Training is the means to bring about continuous improvement in the quality and quantity of work performed by the rural women. Impact assessment relates to assessment after the training is over, on the overall benefits to the rural women as a whole. Methodology used to measure benefits in several occasions is not holistic to serve the purpose effectively. Further, many times the results obtained are not appropriate because of contaminating effects of many intervening factors. Hence, there is need for a matrix mode approach. Narayana's Techniques of Impact Assessment (Fig. 5) in rural women training explains seven attributes like (1) Selection of trainers and trainees i.e. in many rural women training programmes, the selection process is very poor. Hence, proper selection is needed. (2) Assessment of trainer requirements and trainee's needs i.e. many times this assessment is organized in a premature way resulting in poor performance. Hence, proper assessment is very essential. (3) Setting of objectives for rural women training i.e. objectives stated in many programmes are not closely related to the real training needs; very general and unmeasurable objectives

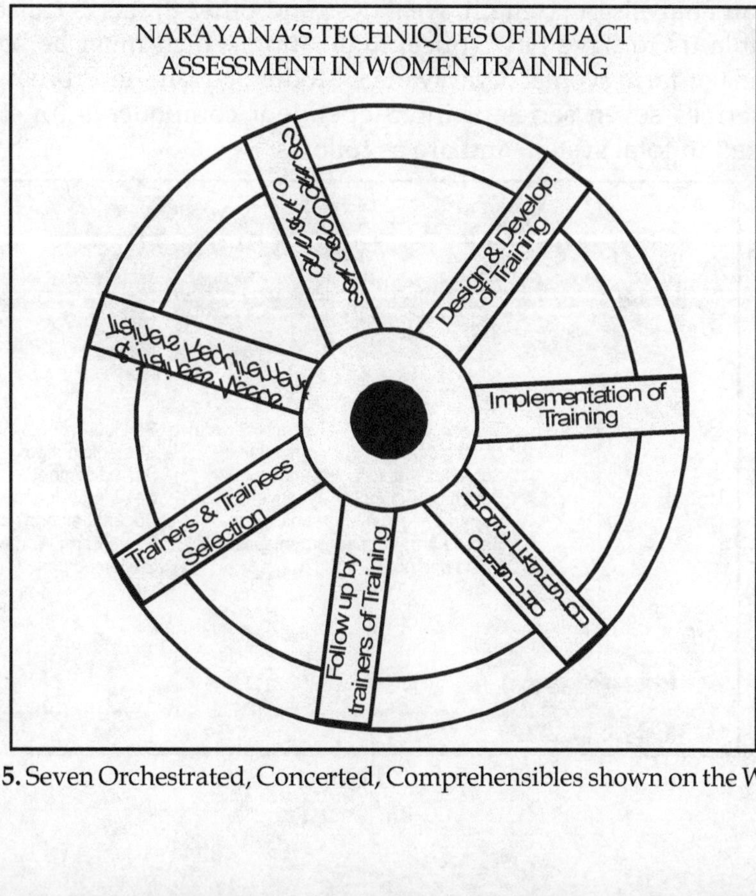

Fig. 5. Seven Orchestrated, Concerted, Comprehensibles shown on the Wheel

leads to poor results. Hence, proper objective setting is a must. (4) The design and development of rural women training i.e. several training programmes designed are not sufficiently related to real training needs or real life situations and circumstances faced by rural women trainees. Hence, it needs proper attention. (5) Implementation of training i.e. well planned rural women training programmes end up with poor implementation in most of cases. Hence, proper execution is important. (6) Impact evaluation of rural women training i.e. this process in several occassions leads to more general and less precision because of poor efforts. Hence, evaluation based on objectives is essential. (7) Follow up by trainers i.e. finding out the utility of learning in the job performance of rural women, analysis of reasons for non utilization of new skills or knowledge acquired, providing health care, reinforcement for wisdom and assessing the future training requirements.

Training of Women Professionals for Agriculture Development: The process of training women professionals must start by questioning the basic assumptions which have governed our training approach so far. So an analysis of strength, weakness and other things is called for to enable us to have new conceptualization which must be based on needs of farm women. Narayana's Model of Training Process (Fig. 6) describes seven series of interdependent components functionally linked to total system and are as follows:

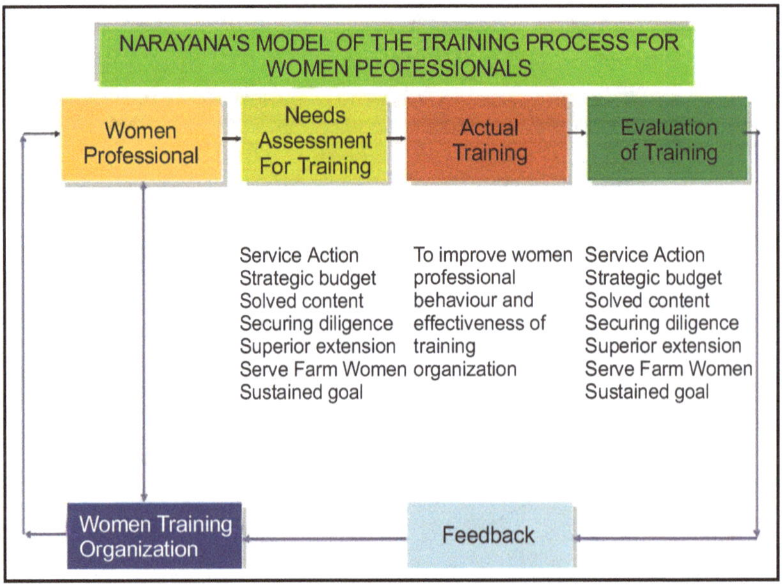

Fig. 6.

(1) *Service action*: Any training must prepare women for service action in the field. If it does not prepare, it needs immediate reorientation.

(2) *Strategic budget*: Many times training strategies overlook the budget provisions. Cost effective training strategies are essential. Hence, use of locally available resources in effective ways is recommended.

(3) *Solved content*: The curriculum designed should help to solve the problems of women. Hence, it requires efforts to develop a training programme with solved content.

(4) *Securing diligence*: Hard work with persistent effort is required for organizing training whether it is need assessment, objective setting, design, implementation or evaluation of training programme. No short-cut method is available for securing better results.

(5) *Superior extension*: Both trainee and the trainer should never forget that the aim of training intervention is to make women to perform better.

(6) *Serve farm women*: The requirements of farm women have to get precedence over professional requirements while training. Our first target should be farm women.

(7) *Sustained goal:* The inner drive, impulse and intention that cause a women to do something in a better way should be developed in a sustained form to satisfy their needs.

Women Resource Development (WRD) as A Challenge for Agriculture Development: During the last five decades, growth of developing countries is directly related to their Human Resource Base. The countries which have given good performance are the countries which have made significant investments in Human Resource Development (HRD). Despite shortcomings in execution of programmes, training continues to secure wide support primarily because of its critical role in HRD. Hence, women Resource Development (WRD) is widely regarded as single most important resource for faster agriculture development.

CONCLUSION

A very important aspect of agriculture development through women is to motivate women for action after recognizing their needs.

Further, sufficient focus is to be given to this aspect in the system. However, experience in the past has indicated that WRD to enlist participation is often lacking. Hence, there is need for training to develop women resource at various levels for agriculture development in India.

23

Dairy Husbandry Training for Rural Women

R.K. PUNIA, D.V.DAHIYA & N.S.YADAV

ABSTRACT

Rural woman plays dominant role in dairy husbandry activities. Capacity building of rural women in this enterprise is the need of the hour to sustain 4% growth rate in dairy sector as the future demand for dairy products is bound to increase. Specific objectives of the study are to understand the socio-economic and dairy husbandry scenario of households; knowledge level about different aspects of dairy husbandry; impact of training on knowledge and; policy implications for their capacity building for this enterprise in the state of Haryana. The data towards this end was collected from a survey conducted on 180 women trainees where pre and post models were used. The background socio-economic variables of women trainees were: 26-45 years age group, literacy of 76.12% and the main occupation was agriculture and household management. Milch buffalo (84.45%) was the main dairy animal. Milk production of more than 5 litres was found in 55.54 percent households and selling of milk more than this quantity was seen in 14.14 percent households. About 60 percent of woman trainees themselves looked after the animals. Pre-training knowledge reveals that majority of women knew the right milking method (70.6%) and breed of buffalo (62.2%). Very poor knowledge was found about the improved practices about space, bringing in heat, mineral mixture, feeding practice and servicing in heat to buffalo. Cumulative knowledge score shows that

44.44, 43.88, 8.33, 3.33 per cent of trainees had poor, medium, good and excellent knowledge level. Post-training knowledge levels were for 2.22, 2.22, 61.66 and 33.88 percent trainees respectively which shows that there was substantial gain in knowledge due to training. Majority of trainees were highly satisfied (77.77%) and perceived the training useful (72.22%). Keeping in view the overall empirical data, secondary data and future vision, there is need to identify potential areas, groups developing specific training modules, creation of infrastructure and delivery system and assured, pre-determined marketing. Rural women need special and specific attention for building their capacity.

INTRODUCTION

Majority of the Asian countries are at an early stage of development, where a high proportion of women are employed in agriculture. Even in most of the South-East Asian countries, where convention prevents women from working outside the home, the care of livestock, processing of harvested crops, crafts and other pursuits provides them means of contributing materially to the family income (Whyte & Whyte, 1982). Women's participation in agriculture (Singh & Punia, 1991) has been significant across the geographical area, cultural boundaries, activities, family systems and development stages. She also contributes through livestock and animal husbandry chores. Haryana is a predominantly agrarian society having the livestock rearing as the second major contributor to farmer's economy with farm women as its main contributor in terms of activities and time, as evidenced by a large number of empirical studies (Achanta, 1982; Kaur, 1987; Vidyalata, 1990; Singh & Punia, 1991; Sikka *et al.*, 2007).

Haryana state being near to national capital and attracting special Economic Zones (SEZs) is marching towards fast changes in agriculture, industries and service sectors. Farm households are bound to be affected by these changes and the most residual economic venture for them would be dairying which their women folk have been doing historically. Kurien (1972), the man behind AMUL (Kaira District Cooperative Milk Producers' Union, Anand, Gujarat) emphasised that dairying is an instrument of change, a means for organizing villages themselves for their betterment; an instrument benefiting both 'haves' and 'have not's; a means of demonstrating modernization and science to the poorest villagers for bettering their own lot and hopeful future of rural families.

Future of rural people depends on their capabilities, skills, entrepreneurship and dynamism in the wake of privatization, globalization and liberalization on all fronts of economic life. Dairy industry will see significant new investment from large corporate houses. To benefit from this, rural people of Haryana need to be equipped through human resource development processes and capacity building strategies. Women constitute about half of the population and economic force. Women and dairying enterprise are integral parts and efforts to train them in scientific dairy husbandry practices (Kuhar, 2004) would lead to rural prosperity. Haryana Livestock Development Board strives for the upliftment of rural masses through various intervention programmes. Training of rural women in dairy animal husbandry was an effort to build capacity of this social segment in this enterprise. Directorate of Extension Education of CCS Haryana Agricultural University, Hisar was given the responsibility of conducting these trainings.

Present paper attempts to ascertain the existing scenario of households regarding dairy husbandry, assess their knowledge about important aspects of dairy husbandry and gain in knowledge due to training, to know their perceived satisfaction and usefulness of training and, to suggest measures for improvements.

METHODOLOGY

Every Krishi Vignana Kendra (KVK) conducted training for a batch of about 30 women in each block in their jurisdiction. Women trainees from seven blocks were selected for the study which constituted the present sample of 180 respondents

Respondent schedule was developed with focused questions and back ground information variables. Senior Extension Specialists of the Directorate, experts from the College and the District were requested to frame the knowledge questions on different aspects of dairy, especially on Black Gold–Murraha buffalo. This knowledge test inventory was carried out before the training and also after the completion of training. Similarly women trainees were asked about their background and other desired variables. Even after the training some trainees were interviewed informally for qualitative responses and reactions for in-depth understanding.

Quantitative data from respondent schedules were edited, compiled and subjected to simple statistical analysis. Qualitative reactions and quantitative data are the basis for interpretation of findings of the study.

RESULTS

Age, education and main occupation of an individual are the most important influencing socio-economic and personal variables for human resource development and our rural development planning. Table 1 given below summaries the information on these variables.

Table 1. Socio-economic background of woman trainees

S.No.	Variable	Categories	Frequency (N=180)	Percentage
1.	Age (Years)	Upto 25	7	3.88
		26-35	58	32.22
		36-45	72	40.00
		46-50	27	15.00
		51 & above	16	8.88
2.	Education (standard)	Illiterate	43	23.88
		Primary	23	12.77
		Middle	21	11.66
		Matric	78	43.33
		Above Matric	15	8.33
3.	Main Occupation (Maximum time (devoted))	Household	35	19.44
		Agriculture	81	45.00
		Labour	17	9.44
		Service & other	47	26.11

Maximum number of trainees was in the potential working age group of 26 to 45 years. They were in responsible position as the female head of family and also a major income earner. Women below 26 years of age were only 3.88 per cent showing that the younger women have different status in the family and only a few were sent by the family for the training. Elderly women (51 & above) constituted only 8.88 per cent of the trainees.

There were only 23.88 per cent illiterate women trainees and 8.33 per cent educated above matriculation. More than half of the trainees were educated up to matriculation and above. Literacy level of the women trainees shows that literates were interested in dairy husbandry training.

Agriculture and household as main occupation constituted 64.44 per cent of the trainees. Though the women reporting household as their main occupation were partially engaged in farming operations and were also performing some chores of animal husbandry. Similarly women with other occupational group i.e. labour and service were also performing some jobs related to animal husbandry.

Current scenario of household: Better planning always needs basic information and existing scenario about the focused developmental programme. Some of the essential information about dairy husbandry of the women trainees is given in Table 2.

In Haryana only 15.55 per cent of households did not have any milch buffalo. Majority of the households had 1 to 2 milch buffalo which is a common practice to meet the milk requirement of the family. More than two buffaloes were in 16.44 per cent of households. Rural households were, thus, having 1 or 2 milking buffalo. Small dairy unit of 4 or more animals was found among 7.22 per cent households of the trainees, which shows that dairying as a subsidiary occupation was expanding.

Table 2. Information about dairy husbandry of households of trainees

S.No.	Variable	Categories	Frequency (N=180)	Percentage
1.	Milch Buffalo (numbers)	0	28	15.55
		1	79	43.88
		2	43	23.88
		3	17	9.44
		4	9	5.00
		5 & more	4	2.22
2.	Milch Cows (numbers)	0	115	63.88
		1	49	27.22
		2	11	6.11
		3 & more	5	2.77

Contd...

Table 2. Contd...

3.	Milk Production (litres)	0	22	12.22
		Upto 2	13	7.22
		2 to 5	43	23.88
		5 to 10	68	37.77
		10 & more	32	17.77
4.	Milk Selling (litres)	0	122	67.77
		Up to 2	3	1.66
		2 to 5	29	16.11
		5 to 10	20	11.11
		10 & more	6	3.33
5.	Care of Animals	Self	107	59.44
		Servant	25	13.88
		Other members	48	26.66

Cow as milch animal was not popular because 63.88 per cent of households did not have even one cow. Three and more number of milch cows were found in household only. Single milch cow met the household needs of milk in the event of not having milch buffalo. Except 12.22 per cent of households, all were having milk production. More than half of the households had more than 5 litres of milk per day. Only in 17.77 per cent households the production was more than 10 litres daily.

Traditionally milk selling was a taboo in rural community but it is no more so with one third of the households selling milk. Among those not selling the milk only 10% did not like and others did not have surplus milk to sell. Participation of women in caring for dairy animals confirms their domination as 59.44% women trainees looked after animals themselves and other members taking care of animals constituted 26.66%. Well to do families and bigger dairy units had hired labourers.

Pre-training knowledge: Knowledge about different aspects of dairy husbandry enterprise develops confidence in rural women, thereby empowering them to take decisions. Before the training, their knowledge about important aspects was assessed by recording their response to structured questions. Pre-training knowledge of the trainees is summarized in table 3.

Table 3. Pre-training knowledge about different aspects of dairy husbandry

S.No.	Aspect of dairy husbandry	Frequency (N-180)	Percentage
1.	Covered space for a buffalo	22	12.2
2.	Milking method by hand	127	70.7
3.	Servicing buffalo in heat	43	23.9
4.	Drying off buffalo before calving	49	27.2
5.	Feeding colostrums to newly born calf	48	26.7
6.	Cutting of navel cord of newly born calf	98	54.4
7.	Bringing animal in heat	15	8.3
8.	Dose of mineral mixture	9	5.0
9.	Breeding buffalo after calving	32	17.8
10.	Feeding of pregnant buffalo	33	18.3
11.	Preventive vaccination	75	41.7
12.	Breed of buffalo	112	62.2

Among the important management aspects, majority (70.6%) knew about the right method of milking by hand followed by cutting of navel cord of newly born calf (54.4%). They did not know the exact space to be provided to a grown buffalo (12.2%). Scientific feeding not only reduces the cost but also maintains the buffalo and its productivity. Dose of mineral mixture considered to be important was not known to 95 percent of the trainees and even feeding of pregnant buffalo was known to only 18 percent trainees. Breed of buffalo–Murraha was known to 62.2 percent women which was due to extensive efforts of extension agencies, specially the state department of Animal Husbandry. In other aspects of breeding i.e. bringing animal in heat, serving of buffalo and breeding buffalo after calving, their knowledge was very poor. Vaccination was known to 41.7 percent of the women trainees.

Knowledge gain is the cognitive aspect and first step in developing capacity through training. Pre and post training knowledge of the trainees reflects the gain in knowledge which is given in Table 4.

Table 4. Gain in knowledge of women trainee's
Knowledge level (Scores) Distribution of respondents

	Pre-training		Post-training	
	Number	**Percentage**	**Number**	**Percentage**
Poor (up to 25%)	80	44.49	4	2.22
Medium (25 to 50%)	79	43.88	4	2.22
Good (50 to 75%)	15	8.33	111	61.66
Excellent (75% & above)	6	3.33	61	33.88
Total	180	100	180	100

Distribution of respondent trainees according to knowledge level before the training reveals that majority (88.37%) was up to medium level which moved to good and excellent (95.54%) category. After the training only 2.22 percent of women remained in poor category i.e. having score up to 25 percent and an equal number in medium category. Majority of women (61.66%) had moved to good knowledge level i.e. score ranging from 50 to 75%. It is concluded that the training has increased their knowledge substantially.

Satisfaction and usefulness: Satisfaction of trainees and their adjudged usefulness of training are indicators of worth and success of the training. Most of the trainees (77.77%) were much satisfied while 16.11% were satisfied and only 6.11% were not satisfied. Similarly, 72.22% judged the training as very useful and 23.33% considered it as useful. In contrast only 4.44% felt it was not useful (Table 5).

Table 5. Satisfaction and usefulness of training

Variable	Category	Frequency (N-180)	Percentage
Satisfaction	a) Much satisfied	140	77.77
	b) Satisfied	29	16.11
	c) Not satisfied	11	6.11
Usefulness	a) Very useful	130	72.22
	b) Useful	42	23.33
	c) Not useful	8	4.44

CONCLUSIONS

Training was imparted to the right target group of rural women as majority was in the age group of 26-45 years having the responsibility of dairy husbandry activities of the households and in economically productive stage. Educated rural women were in majority for dairy husbandry training which is a good indicator for future capacity building in this enterprise. Even the women with service as main occupation were coming forward which may be helpful for better investment in this enterprise. Buffalo was the main milch animal and their number is on the increase, therefore buffalo husbandry should be given special attention. Milk production and milk selling were related aspects which may need the attention of multinational and local entrepreneurs for future developments in dairy industry. There was no hesitation in milk selling also. Training of persons looking after the animals would bring useful results of capacity building. Breeding aspect being the poorest followed by feeding, management and treatment call for immediate remedies. There was substantial gain in knowledge due to training and trainees were satisfied. Thus such trainings are useful and may be proliferated.

The State Department of Animal Husbandry is vigilant and active but there is need to give greater emphasis on capacity building of rural people by imparting such trainings. The future strategy should concentrate on following aspects:

i) Identification of potential areas and groups like peri-agriculture, rural-urban, mini dairy units, commercial dairy farms etc.

ii) Developing training modules for specific groups based on their training need surveys.

iii) Creation of infrastructure and service delivery model to different groups on commercial lines.

iv) Linking the primary producer with market having pre-determined and assured price on sustainable basis

References

Achanta, I.D. 1982. Role of women in agricultural development. *Kurukshetra* 31 (2): 15-17.

Kaur, S. 1987. *Women in Rural Development- A case study*. Mittal Publication, New Delhi.

Kuhar, D. 2004. *Training Needs of Rural Women in Buffalo Rearing Practices.* Unpublished Master's thesis, CCS Haryana Agricultural University, Hisar, India.

Kurien, V. 1972. Dairying as an instrument of change. Rajmitra.A.D. Amin Memorial Lecture Series. Indian Chemical Manufactures Association, National Dairy Development Board, Anand.

Sikka, P., V. B. Dixit, R. K. Sharma & A. Manuj. 2007. Rural women and buffalo rearing. *Kurukshetra* March. Pp. 33-36.

Singh, H. & R. K.Punia. 1991. *Role and status of women in agriculture.* (Ed.) R.K.Punia. *Women in Agriculture*. Vol. I, Northern Book Centre, Delhi. Pp. 303-322.

Vidyalata. 1990. *Developming Rural Women-a case study of Trysem*. Discovery Publishing House, New Delhi.

Whyte, R. O. & Whyte. 1982. *The Women in Rural Asia.* West View Press, Colorado, USA.

24

Strengthening Organic Farming and Food Security Among *Soligas* at Biligiri Rangaswamy Temple Wildlife Sanctuary

SIDDAPPA SETTY, GLADWIN JOSEPH
K.L. RAGHUNANDHAN AND M. GOWDA

ABSTRACT

We are working with *Soliga* farmers and introducing simple 'organic' technologies to increase yields, diversify the farms, and conserve soil and water. Introducing these technologies depends on a strategy of using on-farm trials for low-risk interventions and field-station trials for high-risk interventions. Low-risk interventions include the use of contour row-sowing rather than broadcast sowing, improved access to traditional varieties of seeds, and composting practices. The relatively high-risk interventions are to incorporate locally grafted high-yielding *amla* and related fruit trees, and selecting high-yielding local crop varieties, and crop rotation. All interventions were aimed to strengthen on-farm capacities, improve sustainable traditional technologies and enhancement of livelihood security.

On-farm trials using row cropping showed 30% increase in yields to broadcast sowing. There was a high degree of variation in soil productivity suggesting a potential to improve yields considerably by improving practices and building soils. Traditional varieties of seeds were sourced and systems put in place to store and distribute them. Sixteen tree species preferred by *Soliga* farmers were grown by a few selected farmers and distributed for planting along farm bunds.

We are building on-farm capacity of women and men to improve composting systems. We have recorded 157 different indigenous varieties of seeds and they are being conserved with the help of women in the study area. We conclude that involvement of women in agriculture is very important to strengthen organic farming and to enhance livelihood of the tribal community in Biligiri Rangaswamy Temple Wildlife Sanctuary.

25

Integrated Farming System Approach Towards Efficient Utilisation of Farm Resources by Rural Women

KALPANA, B., G. N. NAGARAJ AND H. KADAR KHAN

ABSTRACT

The green revolution concentrated only on a single enterprise, the crop. Later on, developments in other enterprises like dairy, poultry, piggery, fishery, etc took place, however in an isolated manner. Now it is essential to diversify the economic activities in an interdependent and integrated manner at micro level within the available resources. The system thus developed will be termed as integrated farming system approach. This helps in generating sustained, stable farm income and higher employment. To achieve these objectives, several extension programmers were designed and conducted in integrated farming system demonstration undertaken by the Extension Education Unit of University of Agricultural Sciences, Bangalore in close association with other subject matter specialists. The programmes were implemented in selected *taluks* of Eastern Dry Zone of Karnataka. One such case study is presented here. As a result of continuous technical support and guidance, the farmer was able to increase the net farm income from Rs.48,110 to Rs. 1,39,706 during the final year (2004-05) registering 190 per cent increase over the benchmark year (2001-02).

Hence, it is high time for the scientists, administrators and planners to think on such lines and provide adequate facilities and encouragement to the farmers/farm women to go ahead with the integrated farming systems with a sense of total commitment. This will not only enhance their net farm income and living standard of their families but also reduce hunger and poverty in the society.

SECTION-5
WOMEN CO-OPERATIVES AND SELF HELP GROUPS

26

Women Empowerment through Stree Shakthi Programme An Experiment

VENKATARANGA NAIKA, K. AND
RAGHAVENDRA MURGOD

ABSTRACT

The study was conducted in Tumkur district which is considered as a very progressive district with 7230 SHG's. It has 11 taluks out of which Tumkur taluk was selected for the study. Within this taluk, a cluster of 7 SHG's were considered for the study. The results show that more than three fourths of Stree Shakthi group members came to know about the Stree Shakthi Programme through anganawadi workers (92.5%) followed by news papers (7.5%). It is also interesting to note that 87.5% reported that to a great extent their attitude changed about bringing up their children without discrimination, sending their daughter to school compulsorily and growing vegetables in their backyard. Further results reveal that SSP is changing the attitude of rural setting. Family problems, cast and religion, lack of encouragement in the family, political problem were the major constraints in SHG as perceived by the members. Still SSG has made some qualitative changes in the lives of the rural people.

INTRODUCTION

In India, nearly 50 percent of the female workers are classified as agricultural labourers. About 70 percent of farm work is performed by women. Women's access to land ownership is extremely limited. Majority of rural women depend on agriculture, which is the major unorganized sector in India. Women's contribution to farm sector is around 50-60 percent. They perform two third of the work, but earn only one tenth of the income. Women are lagging behind because of the limited economic opportunity, lack of access for institutional finance and resources, education and minimal participation in decision-making process.

Educational backwardness is regarded as the major reason for women lagging behind and traditionally women lack entrepreneurial ability and marketable skills. As a result, empowerment of women has assumed importance. To raise the status of women, they must be socially, economically, politically empowered. In this direction Karnataka government has launched Stree Shakthi programme (SSP) during 2000-2001 covering 175 taluks. This programme works through formation of women Self Help Groups (SHG's) under the guidance and supervision of Anganawadi workers.

The aims and objectives of the programme are

a. Creating conducive atmosphere to achieve social and financial progress
b. Instilling confidence in self development among women
c. Increasing income of poor women through income generating activities to eradicate poverty
d. Channelising development works to women groups and Extending financial help through lending agencies

The tenth five year plan (2002-07) of Government of India has three fold strategies for empowering women through (a) Social empowerment (b) Economic empowerment (c) Gender issue based on the recommendations of national policy for empowerment of women.

The year "2001" was declared as year of women empowerment by the Government of India. In the same direction the initiative by the

Government of Karnataka to empower women through SHG's is a commendable decision taken by the Government. However, the basic issues are

a) What are the impacts of Stree Shakthi programme on the lives of rural people and knowledge?

b) Are they aware of Stree Shakthi programme?

c) What is their attitude towards the programme and what are the constraints they are facing or suggestions they have to improve the programme?

The specific objectives of the present study are

1. To assess the impact of Stree Shakthi programmes on income generation by women beneficiaries
2. To find out the awareness and attitude of members towards Stree Shakthi programme
3. To elicit the constraints/suggestions to improve the programme

A brief history of women empowerment initiatives in Karnataka and India: Karnataka has a long history for launching empowerment initiatives from 1980's. It is the first state to introduce women training project with the assistance of DANIDA. Government initiatives by both state and central government are projected in table 1.

Table 1. Women's empowerment programmes implemented in Karnataka

S.No	Programme	Year	Government
1	IRDP	1978-79	Government of India
2	Development of women and children in rural area. (DWCRA)	1982	Government of India
3	Rastriya mahila kosha (RMK)	1993	Government of India
4	Indira Mahila Yojana	1995	Government of India
5	Women and Youth training project (WYTEP)	1982	Both state and Central government
6	Swarna Jayanthi Gram Swarojgar Yojana (SGSY)	1999	Government of India
7	Stree Shakthi Programme	2000-01	Governmetn of Karnataka

Contd...

Table 1.Contd...

8	Swashakthi	2001-02	Governmetn of Karnataka
9	Mahila Arthika Swavalambana Yojana (MASLY)	2001-02	Government of India

Salient features of Stree Shakthi (SS) programme and working pattern: In Karnataka SS programme works under the guidance of anganawadi workers up to six months. These workers are school drop outs of X and XII standards engaged in teaching children of weaker sections of society at village level. These workers identify 15-20 women and form the Self Help Group (SHG). Each member has to give Rs. 10/- as membership fee. All the members of SHG were given training on economic activities of government or trained at non-government sponsored training centres at village level or at taluk level. The activities of the SHG will be monitored by the personnel of ministry of women and child welfare department by taluk development officer. At district level, district level co-ordination committee will review the programme under the chairmanship of deputy commissioner.

Table 2. Progress of Stree Shakthi programme as on March 2005-06

1.	No. of SHG'S FORMED	1,00,000.00
2.	Total No. of members	15.12 lakhs
3.	No. of SC members No. of ST members No. of Minorities	3,09,462 1,23,682 1,06,543
4.	No. of groups which opened bank accounts	1,00,000
5.	Total amount savings	348.03 crores
6.	No. of groups which have taken loan from banks	79,022

Table 3. Budget allocation for women empowerment programmes

Year	Amount (lakhs)	Change between 200-01 & 2005-06
2000-01	Rs 832.12	0.28%
2001-02	Rs 1187.70	
2002-03	Rs 1043.41	
2003-04	Rs 3836.16	
2004-05	Rs 1600.00	
2005-06	Rs1168.00	

Source: Ministry of women and child welfare, Government of Karnataka.

Fund: To help SHG's to take up income generating activities government will give Rs 5000 as revolving fund after six months of establishment.

Kit material: Each SHG is given a kit material worth of Rs 600 consisting of register and trunk (metal box to keep books).

Record maintenance: Each SHG will have to maintain Savings account book, loan book, minutes book, receipt book, payment book, credit, cash and debt book, loan account book.

Credit linkage: As per the RBI guidelines nationalized banks will support the SHGs based on their progress.

Credit under SSP is given to the following income-generating activities:

1. Dairying
2. Production & Sale of Garments
3. Blanket marketing
4. Marketing of seeds and manures
5. Production and Sale of *Papad, Agarbathi* and *Sambar* powder (spices)
6. Production and Sale of compost manures
7. Production and Sale of soaps and detergents
8. Production and Sale of cookeries.

Eligibility criteria: The beneficiaries must be

1. Preferably Literate
2. Below poverty line
3. Landless labour or agricultural labourers.

Incentives: If the SHG is able to save Rs. 75000 of their transaction such group will get Rs 15000 as incentive. If it is more than Rs.1,00,000 then Rs. 20,000 is incentive from government.

Rationale: Stree Shakthi programme is based on the firm conviction that socio-economic development of any region or individual depends on the income generation by the individual (women). Further, decision making empowers women under SSP. Inculcating the habit of saving increases the socio-economic status of the member resulting in increased purchasing power of women in general, SSP members in particular. Besides it also develops the entrepreneurial ability of women and results in leadership development among SSP members.

METHODOLOGY

Locale of the study: Stree Shakthi (SS) programme is working in 27 districts of Karnataka covering 175 taluks, of which Tumkur district is considered as very progressive district which has 7230 SHG during 2005-06 and Coorg had least number of SHGs (648). However, based on the number of SHGs, conveyance, distance and nature of activities taken up by the SHGs, Tumkur district was selected for the study. Tumkur district had eleven taluks, of which Tumkur was selected using the said criteria. Within this taluk, a cluster of seven SHGs were considered for the study. Each SHG had its unique names.

Design: Ex post facto

Sampling: The complete list of members of the seven SHGs was obtained, using the list, a sample of 40 women were consulted randomly as sample for the study. All the forty members had taken up diarying as their SSP activity.

Similarly, the Anganawadi workers who were responsible for establishing, supervising and functioning of these these SHGs were also interviewed. Totally 40 SHG members and 10 Anganawadi workers constitute the sample for the study.

Data collection: Data was collected through personal interview method and observation method using a well structured and pre-tested schedule developed for the study during July 2006.

Statistical test: Since the data is descriptive in nature, simple mean, frequency and percentage were used to analyze the data.

RESULTS AND DISCUSSIONS

The results of the study is presented under the following heads:

(a) *Income generation among SHG members:* Table 1 gives a clear picture of income generation among SSG members through their economic activity taken up under SSP. It is interesting to observe that income generation was more after establishing SSG compared to before both in dairying and agriculture. Further, B:C ratio was 1:1.35 and 1:1.31 from dairying and agriculture respectively. This implies the economic activity (dairying) is yielding good results to the members. As a result, the expenditure pattern could be improved, their livelihood could be supported. Increased income is considered as a sign of progress which is found to be true in SSG activity. Thus, SSG member's earnings increased as a result of SSP in the region. This calls for sustained and concerted efforts from both Government personnel and public to keep up the momentum to eradicate poverty in rural areas by empowering women in the family. However, only income generation on sustained basis is considered as a sign of progress.

(b) *Awareness created:* Table 2 indicates that more than three fourth of the Stree Shakthi group members came to know about the SSP through anganawadi workers (92.50%) followed by News paper (7.50%). This indicates that the conventional sources are more credible and easily available, accessible and powerful in creating awareness about the development initiatives.

(c) *Role of slogans:* Table 3 lists the slogans SSP beneficiaries raise in every meeting. It is interesting to note that 87.50 per cent agreed to bring up their children without gender discrimination, 52.5% wanted to send daughters compulsorily to schools and prevent discontinuation of education by daughters. Planting trees and vegetables in the backyard, managing malnutrition, immunising children as per schedule and marrying daughters only after they attained 18 years of ages was acquired to some extent.

(d) *Impact:* It is very curious to note from the table 4 that SSP has made impact to a greater extent on its members relating to increased knowledge about financial sources (100%), developed decision making ability (97.50%), taking up income generating activities (Dairying), book keeping (85%), leadership abilities (80%) followed by confidence in group working (75%) and to some extent social relationship building (75%). Knowledge about programme management, ability of counselling and marketing were developed by SSG members. Thus, it is evident that SSP is contributing to increase in family income of the members besides empowering women to take decisions in the family. As a result not only SSP is helping in personality development of women but also bringing qualitative changes in their lives.

(e) *Constraints*: Table 5 highlights the fact that SSP is not free from problems. However on a personal level, lack of encouragement in the family, family problems and economic problems, marketing were the major problems reported. Joseph & Eswaran (2006) support this finding.

Social problems are very difficult to deal by the motivators or the government personnel because they are very sensitive issues in the community.

(f) *The suggestions* offered by the SHG members are furnished in Table 6. It is evident to note that institutional support (credit) coupled with training and encouragement were the suggestions reported by the majority of the members. This calls for re-thinking of lending procedure in the institutional finance.

Table 1. Income generated by SHG members

S. No.	**Activity**	**Total income(Rs.)**		**B:C Ratio**
		Before	**After**	
1.	Dairying	68500	93000	1:1.35
	Average	1712	2325	
2.	Agriculture	96500	127000	1:1.31
	Average	2412	3175	

Table 2. Awareness of Stree Shakthi group members about Stree Shakthi Programme (n=40)

S. No.	**Source**	**Number**	**Percentage**
1.	Anganawadi workers	37	92.50
2.	Newspaper	3	7.50
3.	Friends	0	0
4.	Family members	0	0
5.	Radio	0	0
6.	Television	0	0

Table 3. Attitude prarthana (prayer slogans) of Stree Shakti Members (n=40)

S. No.	Statements	Greater Extent (%)	Some Extent (%)	Not achieved (%)
1.	We bring up our child in the family without gender discrimination	87.50 (35)	12.50 (5)	0
2.	Send our daughters to school compulsorily and prevent their dropouts	52.50 (21)	47.50 (19)	0
3.	I will plant trees in my back yard	45.00 (18)	55.00 (29)	0
4.	I will manage malnutrition	40.00 (16)	60.00 (24)	0
5.	I will grow vegetables in the back yard	34.50 (15)	65.50 (25)	0
6.	Give immunization at regular intervals	30.00 (12)	70.00 (28)	0
7.	We marry our daughters only after they attain 18 years	17.50 (7)	72.50 (29)	10.00 (4)

Table 4. Impact of SSP on the members (n=40)

S. No.	Statements	Greater Extent (%)	Some Extent (%)	Not achieved (%)
1.	Knowledge about financial sources increased	100.00 (40)	0	0
2.	Developed decision making ability	97.50 (39)	2.50 (1)	0
3.	Took up income generating activities	92.50 (37)	7.50 (3)	0
4.	Learnt about book keeping	85.00 (34)	12.50(5)	2.50 (1)
5.	Developed leadership abilities	80.00 (32)	20.00 (8)	0
6.	Developed confidence in working with group	75.00 (30)	25.00 (10)	0

Contd...

Table 4.Contd...

7.	Gained more knowledge about team building	65.00 (26)	32.50 (13)	2.50 (1)
8.	Developed ability of counselling	42.50 (17)	57.50 (23)	0
9.	Knowledge about programme management increased	27.50 (11)	70.00 (28)	2.50 (1)
10.	Developed social relationship	25.00 (10)	75.00 (30)	0
11.	Developed knowledge on marketing of products	10.00 (4)	52.50 (21)	37.50 (15)

Table 5. Constraints in SSG as perceived by the members (n=40)

Constraints	Number	Per cent
Personal		
1. Family problems	40	100
2. Lack of encouragement in the family	32	80.00
Social		
1. Caste and religion	30	75.00
Political problem	32	80.00
Economical		
1. Lack of capital	30	75.00
2. Lack of security	28	70.0
3. Marketing		
Others		
1. Poor management network for milk products	25	62.50
2. Storage problem	20	50.00

Table 6. Suggestions to improve the functioning of SSG (n=40)

S. No.	Suggestions	Number	Percentage
1.	Banks should give more loans with less interest	40	100.00
2.	Encouragements from husbands	32	80.00
3.	More trainings on value addition	40	100.00
4.	Assured market	35	87.50
5.	More revolving fund	30	75.00

CONCLUSION

It is axiom to note that SSG has made some qualitative changes in the lives of rural people. Majority of the Stree Shakthi group members reported that they came to know (Awareness) about Stree Shakthi programme through anganawadi workers and local news paper. The main sources of motivation for SSG formation were anganawadi workers, friends and neighbours.The changes in the attitude of SSG members were achieved to a greater extent relating to brining up their children, sending them to school, planting trees, managing malnutrition and growing vegetables, whereas the attitude change among SSG members to some extent was observed with regard to getting married their daughters only after 18 years, growing vegetables in their backyard, immunization and managing malnutrition. The impact of SSP on members (women) was observed to a greater extent relating to increased knowledge about financial sources, development of decision making ability, taking income generating activities, book keeping, leadership, confidence in working with the groups, team building in that order.

The major constraints for functioning of SSG were social, economic and marketing. The suggestions for improvement of SSP include more of institutional finance with least interest rate, more encouragement from their family members and more revolving fund from the government.

References

Anon 2003-04. Progress Report, Stree Shakti Programme, Department of Women and Child Welfare, Government of Karnataka, Bangalore.

Joseph. L. & Easwaran. K. 2006. SHG's and Tribal development in Mizoram. *Kurukshetra*. Jan- 2006 pp. 37-47.

27

Performance of Women Self Help Groups Under Swarnajayanthi Grama Swarozgar Yogana (SGSY)

KRISHNAMURTHY, D., B. KALPANA
AND VANITA CHETAN

ABSTRACT

In India, women's economic productivity is rather critical to 60 million households whose family members live below the poverty line. Therefore, the empowerment of poor women is vital for bringing down rural poverty. The operation of the DWCRA project during the VIII plan of government of India (GOI) had inadequate impact and therefore, failed in organizing cohesion within and among the groups to identify the activities that could generate sustained income. Considering the importance of women and lacunae in the Rural development programmes, Swarnajayanti Grama Swarozgar Yojana (SGSY) was evolved and was launched on 1st April, 1999. With this, the erstwhile programmes viz. IRDP, DWCRA, TRYSEM, SITRA, GKY and MWs ceased operation. SGSY aims at establishing a large number of micro enterprises in the rural areas. The 'beneficiary' families (known as Swarozgaris) may be individuals or groups (Self Help Groups). The main objective was to bring every assisted family above the poverty line in three phases subject to the availability of funds. In establishing the micro-enterprises, the emphasis is on the cluster approach.

Keeping this in view, a study was conducted in Bangalore Rural District of Karnataka by surveying the beneficiaries of the selected poverty alleviation programme (SGSY). A total of 120 respondents consisting of beneficiaries from various enterprises, spread over 45 villages and 74 SHGs constituted the sample for the study. The study was conducted with the objectives of measuring the awareness of women beneficiaries about the components of SGSY, the attitude of women beneficiaries towards SGSY, to analyse the impact of SGSY on women beneficiaries in terms of employment generation, and to document the suggestions of women beneficiaries on the content and operational aspects of SGSY. The research study is hoped to be a good source of information to Policy makers, NGOs and Development Departments to empower women through Self Help Groups (SHGs) and to combat poverty in rural areas.

INTRODUCTION

Swarnajayanthi Grama Swarozgar Yojana (SGSY) is one of the poverty alleviation programme sponsored by both Central and State Governments for the benefit of the rural poor living below the poverty line. The scheme was implemented on 1st April 1999 with the definite objective of providing sustainable income generation through micro-enterprise development and also to uplift the rural poor living below the poverty line by creating self-employment.

Awareness refers to the first hand information obtained by the women beneficiaries about the existence of SGSY agency, its activity, its implementation process or its operational procedure. Awareness is very essential because it motivates an individual to obtain further information and to take action. For success of any programme, knowledge on the components, process of implementation and benefits of the programme among the beneficiaries is important. Greater the awareness among the beneficiaries on the components of the programme, higher will be the success of the programme.

This scheme mainly focuses on improving the socio-economic conditions of poor women by encouraging income generating activities through Self Help Groups. To accelerate the over all growth and prosperity of the nation, it was realized that creation of opportunities for socio-economic development of women is very essential.

Accordingly, SGSY emphasizes that organising poor women into SHGs and group activities should be given preference so as to improve their capacity building.

Attitude is the degree of positive or negative feeling associated with some psychological object, person, ideas, institution or a phenomenon. In this study, attitude refers to the degree of positive or negative feelings of women beneficiaries towards Swarnajayanthi Grama Swarozgar Yojana.

SGSY is a credit-cum-subsidy programme. Credit is the critical component in SGSY, subsidy being only an enabling element. It was realized that a sustainable strategy of poverty alleviation has to be based on increasing the productive employment opportunities for women. The programme gives importance to farm women through Self Help Groups. The women beneficiaries are given a list of economic activities such as sheep rearing, dairying, poultry, tailoring, bread making, *venal* making, ready made garments, etc., from which the beneficiaries/Self Help Groups can select any activity for their economic development.

Any developmental programme like SGSY will succeed if the beneficiaries have good opinions about it. Opinions of the beneficiaries provide a means of examining decision making on expenditure and investment of the credit. Opinions and suggestions are also necessary to evaluate any programme and further improve upon its operational aspects. These opinions and suggestions could be taken into consideration while implementing further developmental programmes.

METHODOLOGY

The present study was conducted in Bangalore Rural District based on the fact that maximum number of SHGs were linked to SGSY, since the initiation of the programme. Four taluks were selected for the study namely Devanahalli, Doddaballapur, Hosakote and Ramanagaram. The respondents included women beneficiaries of different SHGs of the above four taluks, and constituted the sample size of 120. A schedule was developed to assess the awareness of the women beneficiaries regarding the SGSY. For this purpose all possible items about the scheme were included and a schedule was developed by referring to literature and consulting experts in the field of implementation of SGSY.

RESULTS & DISCUSSION

Awareness of women beneficiaries on the components of SGSY: Table 1 reveals that all the women beneficiaries (83.3%) were fully aware of the eligibility criteria like presence of their name in Below Poverty Line (BPL) list, membership in SHG, no history of default of other loans and possession of training certificate. These aspects are informed to them during their training period by the training institutes. Also, all of them are completely aware of the documents to be provided by them like Ration or I.D. card, one photo of SHG membership and No due certificate/No objection certificate. None of the beneficiaries were aware of the supporting land document for availing loans. However, it was not insisted by the banks as they take surety of the entire group.

Table 1. Awareness of the women beneficiaries on the components of SGSY (n=120)

S.No	Awareness	Aware		Not aware	
		No.	Percentage	No.	Percentage
I	**General information**				
1	Programme (SGSY)	79	65.83	41	34.17
2	Implementing agency (ZP)	52	43.33	68	66.67
3	**Eligibility criteria**				
	a. Name in the BPL list	120	100.00	0	0.00
	b. Should be member of SHG	120	100.00	0	0.00
	c. Should not be a defaulter	120	100.00	0	0.00
	d. Should posses training certificat	120	83.33	20	16.67
	e. Resident of the village	100	100.00	0	0.00
4	**Documents**				
	Ration card/I.D. card	120	100.00	0	0.00
	One photo of SHG membership	120	100.00	0	0.00
	No due/objection certificate	100	83.33	20	16.67
	Three photoes of SHG representatives	0	0.00	120	100.00
	Supporting land documents	0	0.00	120	100.00
5	Sources of information about SGSY	59	49.17	61	50.83
II	**Credit/subsidy**				
1	Mode of repayment	120	100.00	0	0.00
2	Financial institutions giving loans	120	100.00	0	0.00
3	Period of repayment of loans	120	100.00	0	0.00
4	Insurance policy	100	83.33	20	16.67
5	Rate of interest on loans	100	83.33	20	16.67
6	Amount of eligible subsidy	96	80.00	24	20.00
7	Condition for availing credit	52	43.33	68	66.67
8	Awareness of repayment holiday	0	0.00	120	100.00

Contd...

Table 1. Contd...

III	**Awareness about important economic activities**				
1	Dairy	120	100.00	0	0.00
2	Sheep rearing	120	100.00	0	0.00
3	*Agarbathi*	111	92.50	9	7.50
4	*Venal* making	100	83.33	20	16.67
5	Papad and pickle unit	90	75.00	30	25.00
6	Readymade garments	80	66.67	40	33.33
7	Basket making	76	63.33	44	36.67
8	Tailoring	75	62.50	45	37.50
9	Silk weaving	58	48.33	62	51.66
10	Bamboo products	44	36.67	76	63.33
11	Cocoon garlands	24	20.00	96	80.00
IV	**Implementation of follow up activities**				
1	Training programme	112	93.33	8	6.67
2	Follow up activities	75	62.50	45	37.50
3	Maintaining records	44	36.67	76	63.33
4	Functionaries involved in implementation of the scheme	0	0.00	120	100.00

Regarding awareness about credits/subsidy, all the women beneficiaries (100.0%) were aware of period of repayment of loans, mode of repayment and financial institutions giving loans. The reasons for this could be the intensive contact with the officials of financial institutions lending loans and the effective role played by local leaders, who keep in touch with bank officials. None of the beneficiaries were aware of the repayment holiday. Since, it was not told and encouraged by bank officials, only 43 per cent were aware about all the conditions for availing credit.

Majority of the beneficiaries (100.0%) were aware of the economic activities like dairying, sheep rearing, *venal* making, *agarbathi*, tailoring because these were the activities which are traditionally practiced by women and do not require much training for starting them. Some of the activities like dairy and sheep rearing were also suggested by bank officials since there was no input and marketing problems for the products due to Milk Producer's Co-operatives (MPCs). It was also easy for the banks to recover the loans because of tripart agreement between MPCs, banks and implementing agency.

Majority of women beneficiaries (93.3%) were aware of the training programs conducted by banks as they had participated in the programs in which training about maintaining records of savings and

SHGs were taught. Regarding functionaries involved in the implementation of the scheme, only 62.5 per cent of women were aware of actual functionaries involved. But, most of them could identify bank officials and executive officer since these people are involved in sanctioning of loans. None of the beneficiaries were aware of the follow up activities since they were not conducted frequently.

Attitude of women beneficiaries towards SGSY: The attitude of women beneficiaries towards SGSY is shown in Table 2. Fifty per cent of the beneficiaries had formed favourable attitude towards SGSY and 28.33 per cent of them had highly favourable attitude towards the scheme, but 21.67 per cent had unfavourable attitude towards the programme.

Table 2. Categorisation of women beneficiaries of SGSY based on their attitude towards the programme (n=120)

Category	Frequency	Percentage	Mean (X)	F test value
Unfavourable	26	21.67	52.81	
Favourable	60	50.00	57.13	71.88**
Highly favourable	34	28.33	58.35	

** Significant at 1% level of significance

The data reveals that most of the beneficiaries have favourable attitude towards the scheme; the reason could be due to creation of additional employment to the beneficiaries, which eventually raised their income. It was possible for the women to form such an attitude because all of them have already participated in the programme and most of them are also aware of its components. Some of the objectives of the programme were focussed on up-lifting the rural poor above the poverty line by creating self-employment and income generating activities which also contributed in creating favourable attitude among the beneficiaries.

Impact of SGSY on women beneficiaries in terms of employment generation: The results are summarized in Table 3. A majority (52.5%) of the beneficiaries obtained less number of employment days due to their participation in the programme. Due to their participation, thirty three point three per cent of the beneficiaries could get high number of employment days whereas only a small percentage (14.2%) of the beneficiaries could obtain medium number of employment days from the programme. The results indicate that the programme was successful in creating additional employment for its beneficiaries, since the main aim of scheme was creation of self employment and *micro-enterprises.*

Table 3. Distribution of women beneficiaries of SGSY based on the employment generation

Category of beneficiaries	Number of beneficiaries	Percentage	Mean person days (X)	F test value
Low employment	63	52.50	155.72	
Medium employment	17	14.17	220.34	64.70**
High employment	40	33.33	248.01	

** Significant at 1% level of significance

Opinion of the women beneficiaries towards SGSY: The opinion of the women beneficiaries on the content and operational aspects of SGSY is depicted in Table 4. The women beneficiaries have opined that, the nature of assistance received under SGSY (88.3%), number of beneficiaries selected (65.8%), usefulness of the programmes (77.5%) and the support received from family and society (59.2%) was good and fair. This may be because SGSY was encouraging income generating ventures, which helped the beneficiaries to start their own activities and raise their socio-economic status. They opined that the programmes were very useful to people living below the poverty line. The women also informed that the number of beneficiaries selected was proper because the assistance was not given to each and every person and hence there was no overloading of beneficiaries under SHGs and also the programme. They indicated that there was good support from family and society to participate in the programme because it helped to increase socio-economic status of the families.

Regarding the quantum of assistance received (55.8%), method adopted in identifying the beneficiaries (54.2%), the method of distribution of benefits under SGSY (52.5%) and the support received from banks and NGOs (66.7%), the women beneficiaries indicated favourable opinion, since the beneficiaries under SGSY were identified on the basis of BPL list. This list was prepared by *grama panchayat* well before the implementation of the programme and hence there was less scope for political interruption in identifying the beneficiaries and also the required quantum of assistance was given only to the eligible beneficiaries in kind and not in cash which avoided mystification or diversification of loans, and was utilised for constructive purposes only.

Table 4. Opinions of the women beneficiaries on the content and operational aspects of SGSY

S.No	Statements	Good		Fair		Poor	
		No.	%	No	%	No.	%
1	Nature of assistance received	106	88.33	14	11.67	0	0.00
2	Quantum of assistance received	53	44.17	67	55.83	0	0.00
3	Method adopted in identifying beneficiaries	55	45.83	65	54.17	0	0.00
4	No. of beneficiaries selected	40	33.33	79	65.83	1	0.83
5	Ensuring timely assistance	2	1.67	6	5.00	112	93.33
6	Agency guidance received	0	0.00	3	2.50	117	97.50
7	Follow up visits made	3	2.50	4	3.33	113	94.17
8	Usefulness of the programme	93	77.50	27	22.50	0	0.00
9	Method of distribution of benefits	53	44.17	63	52.50	4	3.33
10	Co-ordination achieved	7	5.83	30	25.00	83	69.17
11	Support received from banks and NGOs	43	35.83	71	59.17	6	5.00
12	Support received from family and society	80	66.67	36	30.00	4	3.33

But, the women beneficiaries expressed dissatisfaction about ensuring timely assistance (93.3%) and guidance by the agency (7.5%), follow-up visits by the officials (97.5%) and coordination between agencies and various line departments (69.2%). This was because there was delay in ensuring assistance and the beneficiaries had to meet the concerned agency officials many times. Several visits had to be made to *panchayat*, while seeking benefits and guidance. Also, there were no follow-up visits by the agency after starting the enterprise and no suggestions were given for further improvement except that some bank officials visited the beneficiaries.

Suggestions of women beneficiaries on the content and operational aspects of SGSY: The suggestions given by the women beneficiaries for the further improvement of the programme are given in Table 5 and discussed as follows.

Suggestions given by the women beneficiaries for improving the SGSY include all SHGs should be graded as good and loans to be provided (100.0%), savings must be ensured in SHGs (56.7%), training must be given about formation of SHG and its norms (83.3%), SHG should be limited, more information should be given on book maintenance by SHG members, loans and subsidy amount must be increased (92.5%), adequate training should be given on some new enterprises (56.7%), provision should be made by the agency for marketing of their products (94.2%), middle man must be eliminated (72.5%), training programmes must be conducted in more numbers in the village itself (99.2%) and duration should be increased (33.3%), panchayat officials must meet the beneficiaries frequently before and after implementation of the programme (95.8%).

Among these suggestions, all the beneficiaries stressed on grading of SHGs as good for sanctioning of loans and amount of loans and subsidy should be increased, provision for marketing of products should be made and training should be conducted on special aspects and follow-up activities to be carried out.

Table 5. Suggestions given by the women beneficiaries for improving the implementation of SGSY

S. No.	Suggestions	No.	%
I	**Formation of SHGs**		
1	All SHGs should be graded as good and loans to be provided	120	100.00
2	Training should be given about formation of SHGs and its norms	100	83.33
3	SGSY should ensure savings in SHGs	68	56.67
4	Group size should be limited	32	26.67
5	More information about book keeping should be given to members	25	20.83
6	More number of SHGs should be established	10	8.33

Contd..

Table 5. Contd..

II	**Loans/subsidy and benefits**		
1	Loan and subsidy amount should be increased	111	92.50
III	**Self employment and income generation**		
1	Training should be given on the enterprise started	68	56.67
IV	**Marketing support ensured**		
1	Provision should be made by the agency for marketing of sheep products, venal, bamboo products and baskets for good price	113	94.17
2	Middle man should be eliminated	87	72.50
3	There should be proper place for marketing of products	8	56.67
V	**Training programmes conducted**		
1	Training programmes should be conducted in the village itself	119	99.17
2	More number of training programmes should be conducted	56	46.67
3	Duration of training must be increased	40	33.33
VI	**Implementation and follow up activities**		
1	Pinhead officials should frequently meet the beneficiaries	115	95.83
2	Follow up activities should be carried out by the agency	88	73.33

28

Constraints Faced and Suggested Measures for Effective Participation and Functioning of Self Help Groups (SHGs)

JAGADEESHWARA, K, T.N. ANAND
A.K.CHAKRAVARTHY AND K.NARAYANA GOWDA

ABSTRACT

The present study was undertaken to analyse factors contributing to participation of rural women in Self Help Groups (SHGs), constraints faced by them and to elicit their opinion to strengthen the role of SHGs. The finding has revealed that, majority of the respondents has suggested that they should be provided with loans and the rate of interest on loans should be less. Further, they also suggested that instead of existing loan structure from the bank and saving amount within the group must be increased which can be invested on other activities. It was also reasoned that the provision of low interest loan from the banks and purchase of raw materials at cheaper rate can improve the financial status of farm women.

INTRODUCTION

The Self-help Group concept is emerging very fast among the rural women, to address the common issues affecting their lives, livelihood, fight against poverty, exploitation and creation of employment generation. It is difficult for rural women to break away from the vicious circle of dependency and poverty. It is only through collective effort and organization that they can reduce their dependency and initiate participation in self-help group activities. Women population in India number 495.73 million. Rural woman constitute 77% of the total population. In order to empower the rural women the government has launched several programmes like DWACRA (Development of Women and Children in Rural Areas) in 1982-83 and Self- help Groups (SHGs) in 2001-02 under WYTEP (Women and youth training for self employment).

To assess the participation level of farm women in economic activities of self-help groups, a study was carried out to know the socio-economic factors contributing to participation level of women in self help groups, to find out constraints faced, elicit the opinion faced for strengthening, effective participation and functioning of self-help groups in Kadur *talk* of Chikkamagalore district of Karnataka State in the year 2004-05.

METHODOLOGY

Kadur *taluk* of Chikkamagalur district in Karnataka was purposely selected because, it comes under southern dry zone where agriculture is having limited scope but there is scope to venture into other avocations. The status of farm women in this selected *taluk* is unique and their participation in agriculture and other agri- based industries are higher. The present study is focused on participation level of farm women in Self-Help groups particularly in Kadur *taluk* of Chikkamagalur district. Eight self-help groups were selected on the basis of total period of existence and the members in each groups available; the total sample size was 120.

RESULTS & DISCUSSION

The major problem faced by farm women to participate in SHG activities was inadequate loan from the banks to start any activity (83.3%). Higher rate of interest discouraged involvement of the members in the activities (66.66%). As their participation was least, they were least interested in saving money on weekly or monthly basis (70.83%). The respondents also said that they were not getting sufficient inputs to organize group activities like provision of separate room and space for demonstratrations (75.00%). It was expressed that lack of cooperation among the members in financial matter (41.66%) was a major constraint. Finally the last but not the least was the non-availability of female extension workers (37.50%).

Table 1. Problems faced by farm women to participate in SHGs activities

S. No	Statements	Numbers	Percent (%)
1.	Provision of Inadequate Loans	110	91.66
2.	High Rate of Interest on Loans	100	83.30
3.	Lack of Inputs to organize the group activities	90	75.00
4.	Inadequate savings from the members of group	85	70.83
5.	More involvement of the members required in all activities	80	66.66
6.	Lack of Co-operation among members in the group	50	41.66
7.	Lack of female extension worker	45	37.50

Table 2. Suggestins made by the farm women for effective functioning of SHGs

S. No	Statements	Numbers	Percent (%)
1.	Bank should provide adequate loans	100	83.30
2.	Provision of bank loans at low interest rate	80	66.66
3.	Membership/savings amount may be increased	75	62.50
4.	Encourage generally interested activities	70	58.33
5.	Appointing lady extension workers to work with SHGs	60	50.00

Majority of them suggested that they should be provided with loans (83.30%) and the rate of interest should be least (66.66%). They suggested that instead of expecting the loan from the bank the saving amount within the group must be increased which will help in investing on many activities (62.50%). They suggested that the farm women should be encouraged to take up general activities along with agricultural activities (58.33%). Finally, they also expressed that lady extension workers should be appointed to work with the SHGs.

29

Participation Level of Farm Women in the Activites of Self Help Groups

JAGADEESHWARA, K, D.M. BASAVARAJAIAH
R. UMA & RAMAPPA PATIL

ABSTRACT

The participation level of farm women in economic activities of self-help groups to know the socio-economic factors contributing to their participation in such groups and the constraints faced by them was studied during 2004-05 at Kadur taluk of Chikkamagalore district of Karnataka. Apart from these findings on the factors affecting their involvement in SHGs and the problems faced by them, the study also revealed that majority of farm women had medium level of family income, mass media participation and, low level of communication status.

INTRODUCTION

Women in India number 495.73 million constituting 49 per cent of total population. Rural women constitute 77 percent of the total women population of the country. Rural women in India are extensively involved in agricultural activities. However, the nature and extent of their participation differs with variations in agriculture production systems. In order to empower the rural women, the Government has launched several programmes like DWACRA

(Development of women and children in Rural Areas) in 1982-83 and Self-help Groups (SHGs) in 2001-02 under WYTEP (women and youth training for self-employment). The Self-help group concept is emerging very fast amongst rural women to address the common issues affecting their lives, livelihood, fight against poverty, exploitation and creation of employment generation.

The Self-Help groups (SHGs) are organizations of rural women, an informal alliance for a common goal to be achieved collectively. In view of accessing the participation level of farm women in the economic activities of Self–Help groups, many studies to know the factors contributing for the participation level of women in the Self-Help groups have been conducted at different places. The present study was conducted at Kadur *taluk* of Chikkamagalur district in Karnataka.

METHODOLOGY

This study was carried out at Kadur *taluk* of Chikkamagalur district, Karnataka. The area comes under southern dry zone where scope for agriculture is limited. Opportunities to venture in to the other avocations are more. The status of farm women in the selected *taluk* is unique and their participation in agriculture and other agri-based industries are more. The focus of the present study was to measure the participation level of farm women in Self-Help groups. Based on the longevity of existence and number of members in a group, eight self-help groups were selected for analysis. The total sample size was 120.

RESULTS

The following facts emerged from the study:

- The respondents were found equally (40%) distributed between old and young age. Majority (58.34%) of them were educated and belonged to the medium sized (68.33%) family with middle level of income. Their participation level in the mass media was only 35.84 percent because of low income. Majority (80 percent) of the respondents fell under low category for communication status. Under the training status 50.83 percent of the farm women came under low category.

- The participation level varied from activity to activity. In the planning activity, majority (40 per cent) of them fell under high category, about 90 per cent of the respondents participated in the group meetings. Only 45 per cent participated in the training programmes. In the implementing activity, majority (62.5 %) of them belonged to high category. In the educational tour activity, 78.33 per cent came under high category. Though the trained respondents were less, their participation level in transfer of technology was high (60.00%). In evaluation and suggestion activity, the respondents belonged to high category (60.83%), and also in the formation of groups the respondents fell under high category (80.66%). (Fig.1)

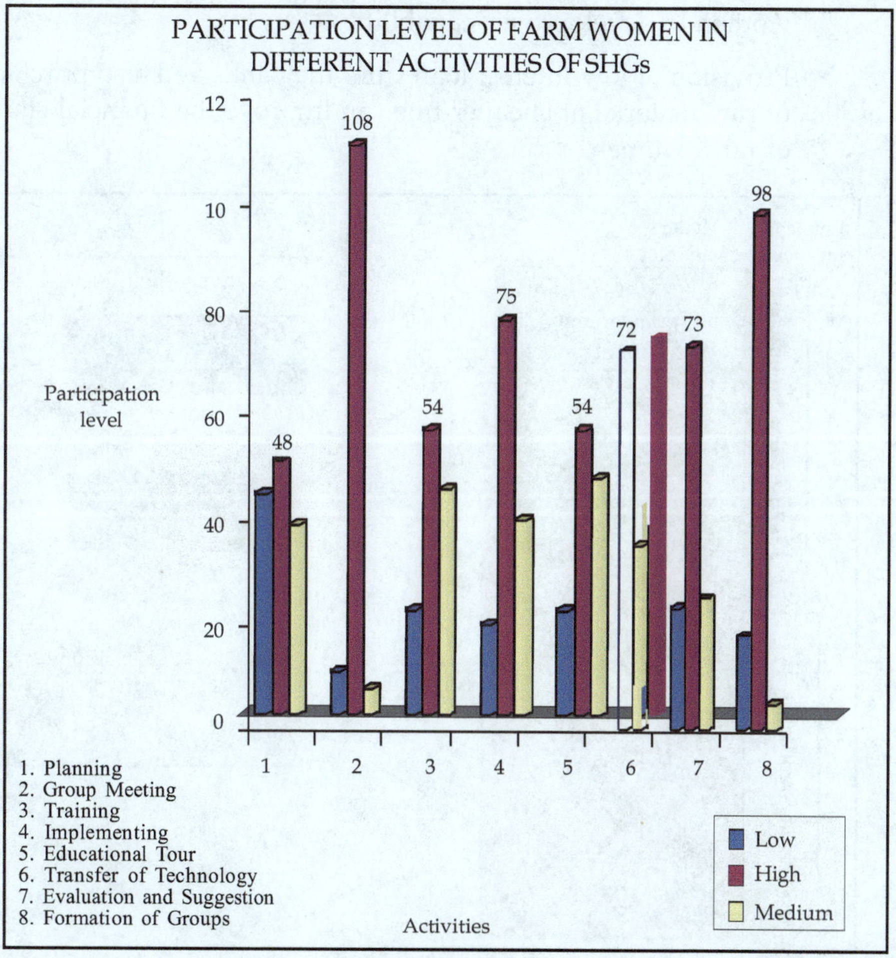

Fig. 1.

- Some of the independent variables like education, training status, mass media participation, communication status, farm status, family size, farm power status, material status were found to be non-significant. Age was negatively correlated with the participation level (Figs. 2).
- Risk orientation, group leadership, empowerment, entrepreneurship behaviour was found to be significantly correlated with the participation level of farm women in all the activities of SHG's (Fig.3). Sufficient exposure to modern technologies through mass media training as well as timely support with inputs was needed for them to perform better. The concerned agencies should take note of this and help the farm women to develop their capabilities.
- Provision of low interest loan from the banks and also purchase of raw material at cheaper rate can improve the financial status of farm women.

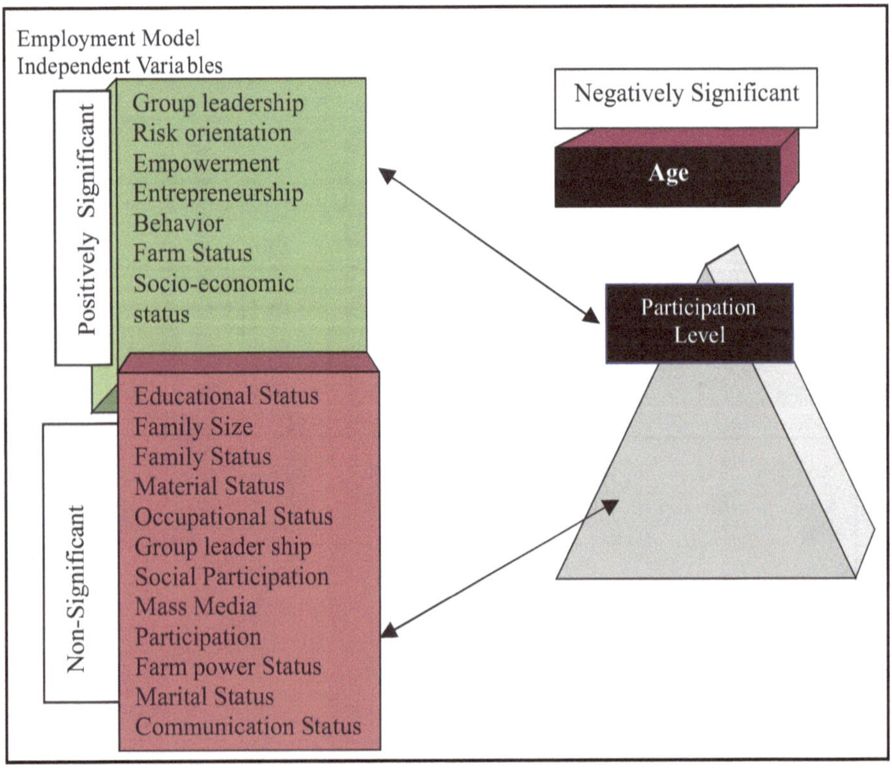

Fig. 2.

DISCUSSION

The findings of the study show that majority of farm women had medium level of family income, mass media participation, low level of communication and training status. It is necessary to consider these variables at all the stages of the self help group activity. The variable material status had found significant relationship in conducting different activities viz., overall participation level, planning stage, evaluation and suggestion stage, and information of groups. Hence, efforts should to be made to improve their material status. This could be possible only when they are given financial as well as technical support. The farm women need to be trained on modern technologies of agriculture, as training showed significant relationship with their participation level. The developmental departments should provide much more training opportunities on various economic activities so that, they can plan employment generating activities. The results strongly indicated that, variables like empowerment of farm women, risk orientation and entrepreneurial behaviour had significant effect on the participation level in group activities. These interrelated variables are very important to develop their overall personality so that they can participate successfully at all stages.

30

Empowerment of Women in Agriculture: Issues and Implications

KALPANA, B., USHA RAVINDRA AND RANI ARVIND

ABSTRACT

Women play a pivotal role in agriculture, in addition to the traditional role of nourishing progeny as mother and attending to the day to day needs of the home as housewife. Women also add to the family income as wage earners. While comprising half of humanity, contributing to two thirds of the world's work hours, a woman earns only one third of the total income and owns less than one tenth of the world's resources. In a country like India which is the home for over 400 million women, the economic status of women is in pathetic condition and they have always been discriminated against men in human resource development with utter disregard to their actual and potential contribution to the national progress in general and agricultural progress in particular. The largest sector of India's economy, agriculture, contributes to 37 percent of the Gross National Product (GNP) and accounts for 60 percent of the employment. The fortunes of over 70 million families directly or indirectly rest on agriculture. Significance of women's role in agricultural development and their contribution in allied fields has long been not only taken for granted but also ignored. There is now a growing realization that we cannot go on ignoring the contribution of nearly half of our population. Things seem to be moving however at a slow speed but in favourable direction. It is important that the speed and the direction of developmental programmes should be evaluated from time to time to ensure that the things are going in the right direction. This paper is an attempt in this direction to analyze the issues and implications of the developmental efforts which have brought lot of positive changes and achievement towards betterment of women folk in rural areas.

SECTION-6
FOOD AND NUTRITION

31

Nutritional Profile and Morbidity Status of Rural Women of Bangalore District

GEETHA. K., G. ESWARAPPA, B.N. MANJUNATHA
B. KALPANA AND C.N. VARSHITHA

ABSTRACT

Women constitute almost half of the population of modern India. Eighty percent of the women population in India live in the rural areas and most of them are engaged in agriculture and allied activities and play a significant role in domestic and socio economic life of the society. Women play a vital role in ensuring proper development and growth of children. They also contribute significantly towards income generation, food preparation and health of the family members. However it is this women segment which is most neglected and suffers from poor health. Health and Nutritional status of this important and substantial segment of the society is very important. The present study investigates the nutritional profile and morbidity status of women in villages around Bangalore city.

INTRODUCTION

Good nutrition fosters sound health, improves the working capacity and facilitates national development. Women's contribution to agricultural production is about 50-60% in India. In spite of the fact that their contribution in agriculture is very high, they are utterly ignored. Several nutritional deficiencies and disorders prevail among rural women. One third of women in Karnataka are suffering from Chronic Energy Deficiency. Hence the present study was taken up to assess the nutritional profile and morbidity status of rural women of Bangalore urban and rural districts.

MATERIAL & METHODS

The study was conducted on 120 randomly selected rural women from four cluster villages of Bangalore District to assess their nutritional and morbidity status. Anthropometric measurements such as weight (Kgs.), height (cms.), mid-upper arm circumference (cms.), triceps fat fold (cms.) waist circumference (cms.) and hip circumference (cms.) were recorded following the standard accepted techniques (Jellifee, 1966). Anthropometric indices such as:

$$\text{Body Mass Index (BMI)} = \frac{\text{Weight (kg)}}{\text{Height}^2\text{ (mts.)}}$$

$$\text{Brokas index (BI)} = \frac{\text{Weight (kg)}}{\text{Height (cm)} - 100} \times 100$$

$$\text{Waist-Hip Ratio} = \frac{\text{Waist Circumference}}{\text{Hip Circumference}}$$

were derived using the measurements based on which the respondents were classified as Normal, Under Nourished or over weight.

Blood haemoglobin level was estimated for Bio-chemical assessment and the subjects were classified as anaemic or non anaemic according to WHO (1975) classification based on the blood Haemoglobin level.

Morbidity Pattern: A good indicator to assess the health status is ascertaining the prevalence of disease such as cold, cough, fever, headache, infection, etc. Episodes of illness 30 days prior to interview were recorded. In addition associated diseases and body conditions were also recorded.

RESULTS & DISCUSSION

Nutritional status is reflected by the body measurements. The mean anthropometric measurements of subjects is presented in Table I.

The results revealed that the mean height and weight was 152.5cms and 49.9kgs respectively which are higher than the values reported for adult females in Karnataka. However the mean weight was slightly lower than the reference weight for Indian adult female which is 50kgs.

The mean Mid Upper arm Circumference (MUAC) met only 80 per cent of the standard for Adult Females (Jelliffe, 1966). The MUAC was found to be higher when compared with data for females from rural areas of Karnataka.

Mean Fat fold at Triceps was 10.99 cms which is lower than that of values reported for rural adult female.

The Mean Waist circumference was 89.3 cms and Hip circumference was 76.4 cms

Anthropometric Indices: The classification of subjects according to anthropometric indices is as shown in Table II. Forty two per cent of the respondents were found to be normal and 18.3% obese and the remaining 39.26 were suffering from chronic energy malnutrition by BMI classification. The findings of the present study with respect to normals by BMI was lower compared with findings of NNMB (1996-1997) according to which only 46% of females were normal by BMI classification.

Waist hip ratio (WHR) classification indicated 41.7% of the subjects to be normal while remaining 58.3 to be obese. However, higher per cent of females were in the category of obesity. Seventy point eight percent of subjects were classified as Normal whereas 5% were overweight and 5% of them were obese according to Brokas Index classification.

As far as arm muscle circumference was concerned the subjects were meeting only 80% of the standard. On the whole, anthropometric indices indicated that majority of the subjects were undernourished and MAC and AMC indicated lower Protein status.

Morbidity Status: Among the subjects suffering from different health related problems, majority of them were found to be suffering from gynaecological problems and headache (Table III). According to Table IV, eight per cent of the respondents were suffering from degenerative disease like Diabetes and five per cent from hypertension. Only 20 per cent of the women under study were having acceptable Haemoglobin level of 12% and above. Hence majority of them were anaemic. (Table V)

Table I. Mean Anthropometric Measurements

Anthropometric Parameters	Measurements	
	Mean	SD
Height (Cm)	152.5	6.4
Weight (Kg)	49.9	8.5
Mid arm circumference (Cm)	24.7	2.8
Triceps fat fold (Cm)	10.99	9.9
Waist circumference (Cm)	89.3	7.7
Hip circumference (Cm)	18.4	7.5

Table II. Respondents by Body Mass Index and Waist Hip Ratio

Anthropometric Indices	Respondents	
	Number	Percent
Body Mass Index (BMI)		
CED Grade-I (16.0-16.9)	10	8.4
CED Grade-II (17.0-18.4)	18	15.0
Low Wt. Normal (18.5 - 19.9)	19	15.8
Normal (20.0 -24.90)	51	42.5
Obese Grade –I (25.0-29.0)	19	15.8
Obese Grade –II (>30.0)	3	2.5
Waist Hip Ratio (WHR)		
Normal (= 0.85)	50	41.7
Obese (>0.85)	70	58.3
Broka's Index (BI)		
< 80 (Under Nutrition)	23	19.2
80-120 (Normal)	85	70.8
120-130 (Over Weight)	6	5.0
130-140 (Obese)	6	5.0
Total	120	100

Table III. Morbidity Status

Aspects	Frequency of Occurrence							
	Daily		Occasionally		Never		Total	
	No.	%	N	%	No.	%	No.	%
Back ache	16	13.3	8	6.7	96	80.0	120	100
Head ache	6	5.0	12	10.0	102	85.0	120	100
Stomach ache	4	3.3	1	0.8	115	95.9	120	100
Gynaecological problems	0	0.0	18	15	102	85	120	100
General Weakness	3	2.5	5	4.2	112	93.3	120	100
Others	6	5.0	6	5.0	108	90.0	120	100

Table IV. Incidence of Degenerative Diseases

Disease	Incidence of Disease (%)
Hypertension	5.0
Diabetes	8.3
Cardiac	0
Renal	0
Others	1.7

Table V. Classification of Respondents by Haemoglobin (%)

Haemoglobin (%)	Respondents	
	Number	Percent
Deficient (7.0-9.9)	70	58.3
Low (10.0-11.9)	26	21.7
Acceptable (>12.0)	24	20.0
Total	120	100

CONCLUSION

In conclusion the present study revealed the anthropometric measurements to be higher than the values reported by NNMB for the Karnataka rural, adult females. Maximum number of subjects were undernourished as judged by different anthropometric indices with majority of them being anaemic. Health related problems and degenerative diseases were also found.

RECOMMENDATIONS & SUGGESTIONS

Education with regard to the importance of health, hygiene and sanitation, micro-nutrients and their importance, low cost nutritious recipies using locally available foods, detection of deficiency diseases and their prevention, value addition to locally available foods such as millets, fruits, vegetables and milk, promoting establishment of nutrition kitchen garden to ensure the accessibility of green leafy vegetables, other vegetables and fruits in their daily diet is required to improve the nutritional and morbidity status of rural women.

References

Jelliffee, D.M. 1966 *Assessment of nutritional status of the community*. WHO Monograph Series No. 53.

WHO 1975. Report of an IAEA/USAID/WHO Joint meeting with specific reference to iron deficiency: Geneva. pp.6-9.

32

Value Addition of Selected Traditional Preparations of Himachal Pradesh

S.R. MALHOTRA, ARCHANA SHARMA AND SATYA BHAMA

ABSTRACT

A survey was conducted in five different villages of Kangra district (HP) namely Padhiarkhar, Lachoon, Avery, Saliana and Patti. Thirty families from each village were selected. A sample of 650 subjects of different age groups was surveyed to assess their food and nutrient intake. It was observed that the food and nutrient intake was less as compared with the Recommended Daily Allowances (RDA). To meet the RDA of the vulnerable section of the society some of the energy rich recipes were prepared under All India Coordinated Research Project in Home Science. The recipes prepared were weaning mix (salty), Weaning mix (sweet), *Papad, Shakerpare, Mathri, Pinni, Amaranth ladoo, Methi ladoo* and *Panjiri*. These products were organoleplically acceptable. No fat content was found in the *papad* but it contained good amount of the vitamin C i.e. 70mg/100g. Maximum fat content was found in the *panjiri* (39.3g/100g). Protein content ranged between 0.78-5.64g/100g, minimum in *panjiri* and maximum in weaning mix (salty). Free fatty acid (oleic acid) content ranged between 33.84-78.96 %, minimum in weaning (sweet) and maximum in *papad*. Peroxide value ranged between 0.95-9.40 percent. These products were prepared by using locally available foods, cereals being the main ingredient. Being nutritionally adequate the preparations were found suitable for infants, school children, lactating and pregnant mothers.

INTRODUCTION

Recipes cooked in Hilly Region of Himachal Pradesh are simple and extremely nutritious. Here one can find huge variety of dals cooked in labour intensive style over slow fire. In Himachal Pradesh staple meals consist of basically *chapaties*, rice along with *dals* (pulses), vegetables and curd (www surfindia.co.). There are also many side dishes like *mathri, shakarpare, panjiri, amaranth ladoo* etc. supplemented with locally available raw materials and consumed as snacks. Diversity in meal pattern can be seen in the local style also. Quite a few of the methods and practices followed by our elders had a very scientific and rational base with major usage of items that are used everyday in the kitchen like milk, turmeric, curd, gram flour, leaves, flower, fruits, seeds, nuts etc. All of these items were widely used not only to beautify but also to maintain and condition the body. Dietary habits of population in different regions of the world vary widely depending upon the local availability of foods but when it comes to the nutritional content of foods, it should be related to cooked foods that reach the table. However, cooking practices vary from one region to another in the country and even from one household to another. So keeping in view the importance of these traditional products, the present study was conducted to standardize and prepare the selected value added traditional food products commonly consumed in Himachal Pradesh.

METHODOLOGY

Information regarding the commonly consumed traditional products, their method of preparation and ingredients used was collected from the housewives of the rural areas of the Kangra District of Himachal Pradesh. On the basis of collected information different preparations were selected and prepared in the laboratory.

Procurement of the samples: Raw materials used for different, frequently consumed regional preparations were procured from the local market. Recipes were prepared and standardized in the laboratory. The prepared recipes were oven dried, ground and kept in air tight containers for further analysis.

Chemical Analysis: Proximate composition of the prepared recipes was done by AOAC (1995) methods. Calcium and iron was analyzed using atomic absorption spectrophotometer (Piper, 1950). Energy content was calculated by multiplication method from protein, fat and carbohydrate contents. Ascorbic acid of the recipes was determined using the method of AOAC (1995) whereas free fatty acids, peroxide value and sugars were estimated by AACC (1976).

Methodology for recipe preparation

Salty porridge

Ingredients: Rice-100g, Green gram *dal*-50g, Amaranth-50g, Salt-2t.sp.

Method: Roast rice, green gram *dhal* and amaranth. Make powder with the help of grinder and mix well. At the time of serving, add little fat and 2 tb.sp. of prepared mixture in a pan and shallow fry it. Add 100ml of water and cook for 2-3 minutes and serve to the child.

Sweet porridge

Ingredients: Wheat flour-100g, Soya flour-25g, Amaranth popped powder-25g, Milk powder-25g, Sugar powder-25g.

Method: Roast wheat flour, Popped amaranth powder and soy flour separately, Cool and then mix with milk powder and sugar powder. Pack in airtight container and store for further use.

Sago papad

Ingredients: *Sago*-100g, Spinach powder-1t.sp, Fenugreek powder-1t.sp, Cumin (jeera) seed -1t.sp, Salt-2t.sp, Water-400ml.

Method: Take *Sago,* spinach powder, fenugreek powder, *jeera,* salt and water in a cooker and cook well for 5-7 minutes. Remove from the fire. Cool it and spread round flat like *chappati* on a tray or polyethylene sheet with the help of ladle. Sun dry it and pack in airtight container for further use.

Shakarpare

Ingredients: Soy flour-50g, Wheat flour-100g, Sugar-80g, Poppy seeds-10g, Coconut powder-20g, *Dalda* (for shortening)-30g, Water-for kneading, Oil -for frying.

Method: Sieve well soy flour and wheat flour. Mix sugar, *dalda* and remaining ingredients well. Knead the flour and make soft dough. Make a thick *chapatti* and cut small diagonal pieces and deep fry till golden brown.

Mathri

Ingredients: Refined flour-100g, Bengal gram flour-50g, Cumin Seeds-2.5g, Salt-1t.sp, Omum seeds-2.5g, Fenugreek seeds-5g, fat - 75g, Water-40ml.

Method: Sieve refined flour, Bengal gram flour and salt. Grind cumin seeds, omum seeds, fenugreek seeds and add to flour mixture. Add oil and make dough using water. Make round balls and flatten it like *chapatti.* Cut it with the help of knife and deep-fry it till golden brown.

Pinni

Ingredients: Soya flour-75g, Bengal gram flour-150g, Groundnut-50g, Sugar -100g, Amaranth powder-25g, Oil-75g.

Method: Sieve soya flour and bengal gram flour. Heat oil in a pan. Shallow fry the soya four and and bengal gram flour. Add groundnut, sugar and amaranth powder and mix well and cook for 1-2 minutes. Make balls and store.

Amaranth ladoo

Ingredients: Soy flour-75g, Amaranth popped-25g, Bengal gram flour-150g, Coconut powder-20g, Cardamom-4nos, Fat-60g, Sugar-100g.

Method: Sieve the Bengal gram flour. Heat oil in a skillet and roast the Bengal gram flour till light brown colour. Mix remaining ingredients thoroughly and make round balls and store.

Methi ladoo

Ingredients: Bengal gram flour-100g, Soya granules-20g, Fenugreek Seeds-20g, Oil -100ml, Sugar-150g, Powdered milk-50g, Desiccated coconut -50g.

Method: Roast the fenugreek seeds and grind finely. Roast soybean flour also. Heat oil in a skillet and shallow fry bengal gram flour, soybean flour and fenugreek powder till golden brown. Add powdered sugar and mix well. Put off the gas, add desiccated coconut and milk powder. Make round balls and store in cool place.

Panjiri

Ingredients: Soya flour-50g, Green gram *dal*-50g, Coconut powder-25g, Raisins-15g, Cardamom-5no., Dry dates-10 no., *Saunf*-5g, Sugar-75g, Fat -40ml

Table 1: Nutrient composition of products per 100g (as is basis)

S. No	Name of product	Moisture (g)	Protein (g)	Fat (g)	CHO (g)	Fiber (g)	Ash (g)	Energy (Kcal)	Iron (mg)	Calcium (mg)	Vitamin C (mg)
1.	Weaning mix (salty)	3.50	15.40	2.00	61.99	1.16	15.95	328	35.5	1.47	40.0
2.	Weaning mix (salty)	2.62	18.55	6.02	68.61	1.10	3.10	403	102.5	0.47	40.0
3.	Papad	6.12	6.65	0.00	81.10	1.08	5.05	315	7.5	0.42	49.7
4.	Shakerpiare	7.68	26.95	17.42	43.62	2.78	1.55	439	187.0	0.86	45.0
5	Mathri	8.94	13.40	24.42	45.78	2.86	4.60	457	16.5	2.30	50.0
6	Pinni	1.08	15.25	25.96	52.29	3.52	1.90	504	54.5	0.53	50.0
7	Amaranth ladoo	2.76	5.25	33.5	56.74	3.86	1.75	329	53.8	1.02	48.6
8	Methi Ladoo	2.86	13.15	36.82	37.28	6.54	3.35	533	33.5	0.67	49.8
9	Panjiri	0.46	14.56	39.30	33.94	8.23	3.46	548	44.5	0.40	48.0

Method: Heat oil in a pan and shallow fry soy flour and green gram dal separately. Pour the ingredients on a plate and spread sugar powder on it, sprinkle the coconut powder and mix well. Store in air tight container.

RESULTS

Nutrient compositions of products per 100g (as is basis) is presented in Table 1. Moisture content of the products ranged between 0.46-8.94 g/100g, minimum in *panjiri* and maximum in *mathri. Papad* showed minimum (6.65g) content of protein whereas *pinni* showed maximum (15.25g) content of protein. *Papad* revealed negligible amount of fat content, whereas *panjiri* revealed maximum (39.30g/100g) fat content. Fiber content ranged between 1.08–8.03 g/100g minimum in *papad* and maximum in *panjiri*. Carbohydrate content varied from 33.95–81.10 g/100g, minimum in *panjiri* and maximum in *papad.* All the products were rich in energy content. Maximum amount of iron (187mg) and calcium (2.30mg) were observed in *shakerpare* and *mathri* respectively whereas good amount of vitamin C was found in all the products.

Table 2: Effect of storage duration on different parameters of products

S.No	Parameters	Storage Duration				
1	Papad	0 days	15 days	30 days	45 days	60 days
	Free fatty acids (% oleic acid)	78.96	79.12	79.26	79.36	79.66
	Peroxide value (meqO2/kg fat)	4.58	5.24	5.75	5.80	5.88
2	Mathri					
	Free fatty acids	33.84	33.89	34.01	34.27	34.37
	Peroxide value (meqO_2/kg fat)	-	0.94	1.20	1.34	1.39
3	Pinni					
	Free fatty acids	50.76	51.09	51.23	52.46	53.16
	Peroxide value (meqO_2/kg fat)	-	3.72	4.21	4.57	5.12
4	Methi ladoo					
	Free fatty acids	67.68	67.98	68.04	68.13	68.15
	Peroxide value (meqO_2/kg fat)	-	2.00	2.07	2.12	2.19
5	Panjiri					
	Free fatty acids	33.84	34.07	34.46	34.82	35.14
	Peroxide value (meqO_2/kg fat)	-	4.35	4.46	4.62	4.79

Effect of storage duration on different parameters of the formulated products is presented in Table 2. The storage duration of the products was 0, 15, 30, 45 and 60 days. Out of nine products only five products were stored to study the effect of storage on the different parameters of the products. The parameters evaluated are free fatty acids, peroxide value, reducing and non- reducing sugars. The free fatty acid content in papad increased with increasing storage duration. The peroxide value of the products was also found to increase with the storage duration. The reducing sugar content was found to be decreased with the increased storage duration. The decrease was 42.8% from 0-60 day's storage duration in *papad*. Non-reducing sugar content was not found in *papad*. The total sugar was found as 3.67 % sucrose at 0 day storage duration. Free fatty acid and peroxide found increased with the increased storage duration i.e. from 0-60 days in *mathri, pinni* and *methi ladoo*. In case of *mathri* reducing sugar was found as 3.17% glucose and the non-reducing sugar was found as 0.81% sucrose at 0 day storage duration with total sugar content being 3.98 on 0 day duration. During storage the reducing sugar content decreased while non reducing sugar and total sugars were increased with increase in storage duration i.e. 0-60 days.

DISCUSSION

While studying the nutritional profile of traditional preparations, a*maranth ladoo* supplemented with soy flour and Bengal gram was found to have good amount of all the nutrients. Manna *et.al.,* (1998) showed that there was a wide variation in physico chemical attributes in amaranth seed based sweetmeats. Sharma *et al.,* (2006) reported that sesame based (*ladoos*) sweet ball formulations were found to be rich in calcium, amaranth based in iron and soy based in potassium. All the nutrients were observed in good amount in *mathri,* weaning mixes and sharkarapare when compared with the nutrient contents of the preparations as given in Nutriguide 2002. The increase in nutrient content may be attributed to incorporation of soyflour, Bengal gram flour, amaranth flour etc. in the different preparations.

CONCLUSION

All the traditional preparations were found to be nutritious and may prove to be helpful in combating the protein energy malnutrition especially in the vulnerable section of the society.

References

AACC 1976. *Approved methods of American Association of Cereals Chemists.* AACC St. Paul, Minnesota, U.S.A.

AOAC 1995. *Official Methods of Analysis.* Association of Official Analytical Chemists, Washington DC.

Manna, J. K., S. G. Kulkarni, H. C. Bisht & I. C. Sharma. 1998. Preparation, packaging and storage of amaranthus seed-based traditional sweet meats. *J. Food Sci.Techn.* 5 (2):161-163.

Nutriguide. 2002. *Manual for calculation of dietary adequacy using nutrient composition of Indian recipes. Technical Bulletin III. Foods and Nutrition.* (Eds.) Tej Verma & Rita Singh

Piper, C.S. 1950. *Soil and Plant analysis.* Inter Science Publication Inc. New York.

Raghuvanshi. All India Coordinated Research Programme in Home Science. Indian Council of Agricultural Research. New Delhi-110 001.

Sharma, H. R., A. Sood, P. Singh, & R. Verma. 2006. Nutrient composition and sensory acceptability of sweet balls based on grain amaranth, sesame and soybean. *J. Dairying Food and Home Sciences* 25 (1): 34-38.

33

Opinion of Scientists About Biotech Foods

PUNIA R. K., SAPNA AND DEEP PUNIA

ABSTRACT

The known views in society about the biotech foods were collected and structured in statements. An opinion survey of sixty scientists of CCS Haryana Agricultural University on these statements was conducted. Half of the scientists agreed that there were no biotech food products in our country. Biotech foods were considered safe for human consumption by 61.7% of the respondents. A large majority of the scientists considered that biotech foods were different in taste from foods made from conventional crops. Similarly 65% of scientists had the opinion that biotech crops increase food allergies. Fifty percent of the scientists agreed that biotech foods were regulated and tested. The rest opined that labeling is needed for biotech foods even in USA. Overwhelming majority felt that biotech foods were not widely accepted. Meat, milk and egg products from biotech animals were not concieved as unsafe to eat by 43.3% respondents. Equal number of scientists felt the same about foods produced through biotechnology and viewed that these were not tested for health. Only one-fourth scientists believed that safety tests by biotech companies were reliable and not self-serving, while the remaining had doubts. Future fear of introduction of new allergens into food supply was perceived by 45% and

33.3% were undecided on this aspect. Difference in nutritional value was not felt by 36.7% respondents. Majority agreed that biotech foods in grocery stores would not be a reality for many years to come. Similar opinion was expressed about acceptance of biotech foods. Overwhelming majority of scientists considered labeling of foods to protect consumers was necessary to promote biotech foods. Majority (55%) of respondents were undecided in giving opinion about transgenic DNA in food about bacteria, cancer and disorders. Policy implication of the study is to make the scientists aware and knowledgeable about the biotech foods.

INTRODUCTION

Global population exceeded six billion at the beginning of century and it is expected to exceed ten billion by 2050. Increasing population in the developing countries of Asia, Africa and Latin Africa may aggravate food insecurity besides the problem of malnutrition. The challenge is to meet the quantitative and qualitative requirements of food at affordable price by doubling the food production from same arable land, maintaining sustainability of productive resources. This phenomenon of achieving food sufficiency and security is an interplay of the complex factors of technology development, dissemination and adoption, social system in which new foods are introduced and accepted, political acceptance and will at national and international level.

Technology development, its dissemination and adoption have played important role in food production and productivity. But the process had been very slow till 1960s, when the new varieties of rice and wheat responsive to input factors especially the fertilizer and water were introduced. Rice production increased from 30.4 million tones in 1966 to 93.3 million tones in 2001-02 and maximum productivity gain was in 1970s and 80s. It could not be maintained from 90s onwards. Instead there is a plateau. Similar trend is seen in wheat production and productivity. There is little scope for increasing area under the crops to increase production and productivity without a breakthrough in genetic potential. The conventional techniques of selection and breeding may not be sufficient to break this stalemate in productivity.

Biotechnology seems to be panacea by developing genetically modified (GM) crop varieties which have resistance to drought, waterlog, soil acidity, salinity, extreme temperatures, insect-pests and diseases, herbicides, improved nutritional qualities, reduced allergincity, enhanced shell life and increased plant productivity. This non-conventional technology has the ability to move desirable genes within and across to create new varieties. It provides access to a greater gene pool as the gene(s) may come from viruses, bacteria, human beings, fungi, insects, animals, unrelated plants and even from chemical synthesis in the laboratory. India has developed transgenic crops like rice, wheat, black-gram, chick pea, pigeon pea, sorghum, brassica, ground nut, sunflower, cotton, sugarcane, brinjal, cabbage, cauliflower, musk melon, okra, potato, tomato, water melon and tobacco. In other parts of the worlds, GM crops of soybean, maize, canola, squash and papaya have also been developed. Of the 159 field releases of GM crops, Argentina leads with 43, followed by Chile, Mexico, Puerto Rico and Republic of South Africa with 17-20 each. India is far behind with just two field releases (de Kathen, 1996). Area under transgenic crops has increased from 1.7 million hectares in 1996 to 27.8 million hectares in 1998 (James,1998) which has further increased to 52.6 million hectares in 2001 and proliferated to 90 million hectares by 2005. In India the adoption of Bt. cotton has increased from 0.3 lakh hectares in 2002-03 to 30 lakh hectares in 2006-07, occupying 33 percent of the total area under cotton crop in the country. Commercialization of transgenic crops is being adopted at a very fast rate.

Transgenic crops do not find ready acceptance by the farmers and the end-users of produce, the consumers. Bhat & Chopra (2005) mention that "Bio-safety" concerns are the major reasons for non-adoption of transgenic crops in most countries. These concerns include transgenic movement to other varieties and wild relatives leading to possible development of super weeds, eroding of genetic diversity and ecological disturbances. Similarly, there is widespread apprehension that transgenic products could be toxic or allergic to humans and animals. Also, transgenic crops (e.g. pest-resistant varieties) could have adverse impact on non-target organisms. Emergence of more virulent forms of pests and pathogens is also an important concern. It is this ability of gene movement within and across the living beings that makes it not only potentially powerful tool for modifying nature but also raises ethical, health and environmental issues which are being taken up by National and International Organizations against biotechnology/GM crops (http.:/ www.agbioresources.com – A Resource Website on Agricultural Biotechnology – Designed and maintained by Srinath). Many concerns raised about agricultural biotechnology may be genuine but these lack scientific validity. There are several beliefs about biotech foods leading to reservations for not consuming these non-conventionally produced foods. Public opinion and propaganda have also contributed to the generation of such myths in the society. Unless these myths are dispelled, the scope of biotechnology will remain limited and the objective of the second green revolution by increasing productivity will have limitations.

The present empirical study is an attempt to ascertain the opinion of scientists about the popular myths regarding biotech foods and suggest some measures for dispelling these myths for better acceptance of biotech foods.

METHODOLOGY

Food is a very sensitive issue where in social and cultural cues play very important role in its acceptance or rejection. Food fads, myths, beliefs, totem and taboos influence opinion formation about it. Biotech foods are almost unknown to general public. Therefore a first hand opinion of sixty scientists of this university was gathered. Moreover the scientists of an agricultural university are assumed to be the most informed persons about the biotech foods.

Literature on sociology of food was reviewed and the misconceptions, beliefs, taboos, fads, myths etc. were enumerated regarding biotech foods. Information from common citizens was also collected. The most comprehensive views in this area were presented by Bhatt & Tomar (2006) which were extensively used after modifying to formulate opinion statements. Questionnaire containing the structured statements was presented to scientists to give their opinion on a five point continuum i.e. strongly agree, agree, undecided, disagree and strongly disagree. Frequency response for each statement was calculated for interpretation.

RESULTS

Opinion statements were broadly categorized in six areas and responses of scientists are given as under.

1. *Availability of biotech foods* in the market for human consumption, was agreed upon by half (50%) of the respondents whereas other

half were not aware about biotech food products in the market. Less than half (43.3%) agreed that there were no such foods in the market and 6.7 per cent were undecided and could not reply candidly.

The opinion of 53.4 percent scientists was that presence of biotech foods in grocery stores may take many years. Fast development in this line was opined by 26.7 percent only. Acceptance of biotech foods in the society was favorably viewed by 13.4 percent while the majority of scientist respondents (75%) stated non-acceptance.

2. *Belief about safety* to human beings in eating biotech food was held by 61.7 percent and unsafe to eat by 8.3 per cent only. About one-third (30%) were undecided on this important aspect.

The generally held myth that GM crops are not safe was held by only 11.6 percent of scientists. Half of the respondents were of the opinion that the GM crops were safe and 38.3 percent were undecided about this important aspect.

The belief that biotech crops/foods increase allergies, was held by 26.7 per cent and 38.3 per cent were undecided. Almost the same number of scientists opined that the biotech foods did not increase food allergies.

Majority (53.3%) of the respondents had the general view that meat, milk and eggs from livestock and poultry fed on biotech feed products were safe. Only 15 per cent did not feel the byproduct safe and 31.7 per cent did not have decisive response.The indirect effect was lesser.

The opinion that growing drugs in plants is dangerous was held by 38.4 percent scientists followed indecision by 35% and disagreement was expressed by 26.7%.

One-fourth of the respondents opined that meat, milk, and egg products from biotech animals are unsafe for human consumption, whereas 43.3 percent did not perceive it unsafe. About one-third (31.7%) were undecided.

Fear of putting susceptible people at risk due to new allergens introduced in food chain by biotech techniques was agreed by 45 percent respondents, followed by undecided (33.3%) and disagreed (21.6%) scientists.

The commonly held opinion that transgenic DNA may cause disorders in body was not decisively agreed by 55 percent scientists. About 26.7% agreed it to be so while 18.3 % disagreed. Forty percent of respondents thought modern biotech being inherently a different process from conventional breeding poses greater risks. Only 21.7 per cent did not believe so. As many as 38.3 percent respondents were undecided about this myth.

The belief that GM involves incorporation of dangerous gene products into crops was believed by 23.4 per cent but 36.6 per cent did not perceive so. A higher number (40%) did not decide either way.

3. *Differentiation between the biotech and conventional foods* was perceived by about half (48.4%) of the respondents who believed that the taste of biotech foods was different than foods made from conventional crops. Only 20 per cent disagreed that there was any difference in taste and, 31.7 per cent were undecided.

   Biotechnology for improving the plants as un-natural process/ technique was agreed upon by majority (63.4%) and disagreed by 28.3 per cent of the respondents.

   Opinion that transgenic and cloned animals are different from normal animals was agreed by about half of scientists. About one-fifth were undecided and one-third (30%) disagreed to this opinion statement.

   Differential nutritional value of the food was perceived by 36.7 per cent respondents and an equal number of respondents (36.7%) did not perceive the difference. About one-fourth could not decide about the nutritional difference between biotech and conventional foods.

   The notion that biotech foods contain genes that are not present in conventionally grown foods was unacceptable to 48.4 percent scientists while 33.3 percent were undecided and 18.3 percent agreed. More than half (56.7%) of the scientists were undecided about the generally held view that transgenic DNA in food is taken up by bacteria in human gut. One-fourth of them believed so and 18 per cent did not.

   From the nutritional differentiation point of view, majority (56.6%) believed that food derived from organic or conventional crop was better than food from biotech crop. About an equal number of scientists disagreed with this superiority of conventional food and an equal number were undecided about this differentiation.

4. *Requirement of labeling* of biotech foods is an essential aspect. Since the United States leads in developing biotech foods, the questionnaire sought information about the status of labeling there. Only 5 percent scientists responded that US did not require labelling, whereas majority (55%) were for mandatory labeling even in USA. Forty per cent responding were undecided.

   Overwhelming majority (85%) of scientists responded that generally consumers want labeling of biotech foods while 6.7 percent did not affirm it and 8.3 percent were undecided.

   Lack of regulation for biotech food was perceived by one-third of scientists. Only 23.3 percent perceived any regulation and 43.3 percent had undecided opinion. Half of the scientists responded that biotech foods are regulated or tested, other half disagreed or undecided to this important aspect of food safety.

   The notion that foods produced through biotechnology are not tested for health and environmental safety was agreed by 31.7 percent scientists and relatively more (43.3%) disagreed. One-fourth of scientists did not respond in a deciding way about this aspect

   The general view that safety tests of biotech foods conducted or sponsored by biotech companies are unreliable and self-serving was held by 35 percent of scientists and 26.7 percent did not hold this opinion. Thirty eight point seven percent did not have a candid opinion. Labeling of biotech foods to protect consumers and increasing consumer choice was the perceived requirement by overwhelming majority (88.3 %) of interviewed scientists.

5. *View on welfare of* animal about pain or distress was held by 23.4 percent scientists and about an equal number denied it. About half (48.3%) of the respondents did not decide candidly.

   Environmental concern was perceived by 23.4 percent and 36.7 percent did not feel so. Larger percentage (40%) was undecided about the effects of biotech crops on environment.

Increasing use of biotech for producing pharmaceuticals and drugs through food crops was acceptable to half of the scientists. This innovation was not agreeable to 18.4 percent and 31.3 percent were undecided.

Half of the respondents believed that the consumers are opposed to biotechnology in food production and only one-fifth (21.6%) did not perceive such opposition from consumers.

6. *Promised benefits of biotechnology:* It was agreed by 41.7 percent scientists that GM crops failed to deliver their promised benefits followed by undecided (31.7%) and disagreed (26.7%) scientists.

   Majority (60%) of scientists were of the opinion that agricultural biotechnology will benefit developing countries. Only one-fifth agreed with the general notion that it will not benefit the developing nations.

   The claim that yields of herbicide tolerant soybeans are lower than conventionally bred soybeans could not be decided by 55% of the scientists. Only 18.3 percent disagreed with it and 26.6% opined affirmatively.

   Potentiality of biotech foods to feed the world was accepted by only 28.4 percent and a relatively larger (41.6%) number opined otherwise. About one-third (30%) were undecided about the potentiality of biotechnology to produce sufficient food to feed the world population.

CONCLUSION

Opinion of the scientists about any commonly held belief or myth is always influenced by his knowledge about that aspect. Study data clearly shows that a fairly good number (>25%) of them were undecided on safety aspects of biotech foods, technical matters and regulatory provisions. Even on safety matters they were in variance with the general opinion. Their views were also in congruence with generally held opinion in respect of consumer aspects and acceptance of biotech foods. Regulatory measures were neither clear nor defined which is the most important aspect for acceptance of biotech foods in the elite society at the initial stage. People still believe that organic foods and foods produced through conventional methods are better than biotech foods.

SUGGESTIONS

Awareness and education are the tools to dispel the beliefs formed in public by the anti-biotechnology groups through social movements and media. De-freezing of public myths is the first step and scientists can provide the answers to their doubts and questions raised. Anti-biotech lobbies may be made knowledgeable in the desired right perspective followed by enlightenment of public.

Agricultural biotechnology primarily deals in food production and food is a very sensitive issue of personal concern of the public. Therefore, the public must be satisfied about the safety of food, safety of the environment, the effectiveness of the regulatory measures, and relative advantage of technology. The consumers and the public should be provided with realistic and complete information truthfully for understanding the benefits and limitations of technology.

Pro-technology groups, multinational corporations and the governments should be objective and truthful in their dealings of providing correct information, enforcing regulatory measures and owning responsibility. The confidence of public by making them aware and educated is necessary for acceptance of biotech foods. They should also bear the financial burden of public awareness and education programmes.

References

Bhat, D. K. & Tomar, Priyanka. 2006. Biotech foods, myths, misconceptions, misinformation and facts. *Processed Food Industry* 9(4) Feb.2006. pp.36-45.

Bhatt, S. R. & V. L.Chopra. 2005. Transgenic crops: Priorities and strategies for India. *Curr. Sci.* 88(6):886-889.

De Kathen, A. 1996. The impact of transgenic crops in developing countries. *Biotechnology and Development Monitor* 28:10-14.

James, C. 1998. *Global Review of Commercial Transgenic Crops.* 1998. Ithaca, New York: International Service for Acquisition of Agri-biotech Application, pp43.

34

Value Addition to Local Farm Produce Through Rural Women SHGs and It's Economic Viability

ESWARAPPA, G., K. GEETHA, B.N. MANJUNATHA AND C.N. VARSHITHA

ABSTRACT

Self Help Group (SHG) has emerged as a new paradigm for combating poverty and self help promotion has become one of the major developmental strategy in the country today. The self help groups are motivated to take up suitable micro entrepreneurship for income generation either individually or in groups, which is dependent on the resource availability and their skills. The empowered SHG women have started different small enterprises such as *papad* making, malt preparation, different value added products, and other agro based industries which has increased their income. In recent years agro processing and value addition to the farm produce have attained great socio economic significance. A massive thrust to food processing and other agro based industries will add value to the product there by increasing the income of farmers, create employment opportunities, diversify rural economy and foster rural industrialization. Hence the present study was undertaken to access the economic viability of the self help groups involved in value addition to *Ragi*, a finger millet. *Ragi* malt, *Ragi papad*, *Ragi unite* (Energy food) and *Ragi* flour were the value added products prepared by these SHG's. The self help groups of Chintamani and Anekal taluk were selected for the study. The results indicated that value addition to *Ragi* has not only created job and income for rural women but also increased the value (selling cost) of *Ragi* thereby empowering women SHG members both socially and economically.

35

Women and Food Security in South Asia : Current Issues and Emerging Concerns

NIRA RAMACHANDRAN

ABSTRACT

The food security scenario in South Asia has witnessed rapid progress over the last few decades, yet nutrition outcomes, especially those related to women and children, have failed to keep pace. This paper contends that the role of women in providing food and nutrition security at the household and individual level needs to be examined, if the paradox of persisting malnutrition amid macro level food sufficiency is to be resolved. Food security, in its broader connotation, results from the availability of adequate food, effective consumption, and desirable nutrition outcomes. As such, it is intricately linked with a woman's multiple roles expressed in her productive, reproductive, and caring functions. However, even focused efforts aimed at resolving the problems faced by women in performing one or other of their roles, may fail to produce expected results, if the issues underlying each function and their inter-linkages are not fully understood. The paper thus attempts to review various aspects of the relationship between women and food security in South Asia, highlights the issues that require urgent focus and indicates emerging concerns in the region.

The educated and socially empowered Asian women is the key to improving the nutrition and mental acuity of young children and that improvement sets in motion lifelong prospects for heightened learning and earning with benefit streams to families.

*This paper was prepared for the UNU-WIDER project on Hunger and Food security: New Challenges and New opportunities, directed by Basudeb Guha-Khasnobis. The project was carried out in collaboration with the Indian Council of Social Science Research (ICSSR) and the UN Food and Agriculture Organization (FAO).

UNU-WIDER gratefully acknowledges the financial contributions to its research programme by the governments of Denmark (Royal Ministry of Foreign Affairs), Finland (Ministry for Foreign Affairs), Norway (Royal Ministry of Foreign Affairs), Sweden (Swedish International Development Cooperation Agency-Sida) and the United Kingdom (Department for International Development).

ISSN 1810-2611 ISBN 92-9190-915-7 (internet version)

**UNICEF, New Delhi ; email: nira_ramachandran@hotmail.com

INTRODUCTION

The last few decades have witnessed a sweeping change in the food production scenario of South Asia. India, the largest of the South Asian economies, is now largely self-sufficient in food grain and an emerging exporter. While the other countries of the region remain dependent on cereal imports, the per capita availability of cereal has increased in every country (with the exception of the Maldives) from the 1980s to date (FAO 2002). Yet endemic pockets of hunger remain, seasonal shortfalls are manifest and malnutrition is widespread across the region, women and children being the greatest sufferers. The 'Asian enigma', as it is termed, of food scarcity and malnutrition amidst plenty, has defied all attempts at resolution so far. Poverty alleviation strategies, livelihood generation programmes and direct food interventions have all been tried, to little avail. Food security researchers have often commented on the fact that, while most South Asian countries have available food stocks and better health and education services in comparison to many other developing countries, even most countries of food-deficit Sub-Saharan Africa (SSA) score higher in terms of the nutrition levels of their women and children. It is thus evident that a fresh approach to the issue is warranted, if the millennium goal of a hunger-free world by 2015 is to be achieved.

Over the past decade, international agencies have made efforts to internalize gender concerns in all development issues. In many areas such as education, health and micro-finance, for example, impressive results have been achieved with this approach. The issue of women and food security, however, which has also received similar attention, has failed to translate into the expected results. The slow-paced response to gender-based food security efforts reflects the complexity of the relationship between the two. Food security, in its broader connotation, results from the availability of adequate food at country level, household and individual access to adequate and nutritious food, effective consumption and adequate nutrition outcomes-all in a sustained manner. As such, it is intricately linked with woman's multiple roles expressed in her productive, reproductive and caring functions. Even focused efforts aimed at resolving the problems faced by women in performing one or other of their roles, may fail to produce adequate results if the issues underlying each function and their inter-linkages are not fully understood.

This paper attempts to review the various aspects of the relationship between women and food security in South Asia, highlights the issues requiring urgent focus and indicates emerging areas of concern. The paper is organized in nine sections. Section 1 provides the background to the issue; section 2 deals with food security, nutrition and health indicators in the region and draws international comparisons; section 3 discusses the issues faced by the woman farmer; sections 4 and 5 bring out gender disparities in livelihood options and differential wage structures, while section 6 highlights the positive nutritional impacts of women's contribution to household income. Section 7 reveals the existence of gender disparities in food security within households, while Section 8 focuses on the inter-generational consequences of unequal access to food and healthcare. The last section summarizes the findings of the review.

THE SOUTH ASIAN REGION: INTER AND INTRA-REGIONAL COMPARISONS

As compared to other developing regions, South Asia ranks below average with respect to several human development indicators (HDI). Table 1 brings out the position of the region vis a vis other developing regions with reference to our selected indicators reflecting food security, health education and the combined impact of these as mirrored in life expectancy at birth. It is clear that the region ranks only above SSA and well below developing-country average in all aspects.

In fact, country-level comparisons reveal an even more dismal picture with the countries of South Asia consistently recording lower values on nutrition, health and education indicators, particularly with reference to women, than many of those in SSA (Table 2). A point to be noted is that all the countries selected from SSA rank lower on the HDI than those from South Asia, yet women and child specific nutrition indicators such as infants born with low birth weights and the percentage of undernourished children below five years are consistently better in the former region. Even income poverty indicators fail to adequately explain the difference in food and nutrition security. Countries with similar poverty ratios score very differently on child nutritional outcomes. In every case (with the exception of Ethiopia), the South Asian countries exhibit lower scores.

That South Asia's poor showing in human development indicators can be traced to the vicious cycle of malnutrition, ill-health, illiteracy and poverty which persists in the region is well-known. That globalization and structural adjustments have led to an economic upswing at macro level in most of South Asia, but the observation that disparities in incomes, livelihoods and human development are accentuating is also accepted. However, the realization that the roots of the problem lie in the gender discrimination, which is prevalent in most of South Asia, is only now gaining credence.

Table 1. Inter-regional Comparisons

Region	Life expectancy at birth (yrs) 2002	Underno-urished people as % of total population 1999-01	Infant mortality rate (per '000 live births)	Adult literacy rate (% of 15-yrolds and above), 2002
All developing countries	64.6	17	61	76.7
East Asia & the Pacific	69.8	–	32	90.3
Latin America & the Caribbean	70.5	11	27	88.6
South Asia	63.2	22	69	57.6
Sub-Saharan Africa	46.3	32	108	63.2

Source : UNDP (2004).

Referring to India's malnutrition problems, Nobel Laureate Amartya Sen (online, undated) writes, 'Since maternal under nourishment is causally linked with gender bias against women in general, it appears that the penalty India pays by being unfair to women hits all Indians, boys as well as girls and men as well as women'.The statement could well apply to the entire South Asian region.

Strengthening this viewpoint is the one exception to South Asia's poor food security and nutrition record, that is, Sri Lanka. Sri Lanka has long been cited as a role model for the developing world. With a GDP at developing-country level, yet it has attained social indicators comparable to those of the developed world. Part of its success may be attributed to specific feeding programmes and an early emphasis on universal education, but a large share of its success is, perhaps, attributable to the overall gender equity in the country.

Table 2. Inter-country comparisons

HDI Rank	Country	Children under- weight for age (<5 yrs),1995-00	% Low birth weight inf- ants, 98-02	Maternal mortality rate 2000	Female adult literacy rate, 2002	% Population living below $1 per/day
Sub-Saharan Africa: selected countries						
166	Angola	31	12	1700	-	31
161	Benin	23	16	850	25.5	23
167	Chad	28	17	1100	37.5	28
163	Cote d I'voire	21				
170	Ethiopia	47	15	850	33.8	47
148	Kenya	21	11	1000	78.5	21
145	Lesotho	18	14	550	90.3	18
131	Ghana	25	11	540	65.9	25
165	Malawi	25	16	1800	59.2	25
146	Uganda	23	12	880	-	23
South Asia						
138	Bangladesh	48	30	380	31.4	48
134	Bhutan	19	15	420	-	19
127	India	47	30	540	46.4	47
140	Nepal	48	21	740	26.4	48
142	Pakistan	38	19	500	28.5	38
96	Sri Lanka	29	22	92	89.6	29
84	Maldives	30	22	110	97.2	30

Source: UNDP (2004)

LAND RIGHTS AND THE WOMEN FARMERS

A large portion of the world's food output originates in the hands of women farmers. In most developing countries, women provide over half the agricultural workforce. Across South Asia, however, women, on an average, account for about 39 per cent of the agricultural workforce. Country-wise variations range from 50 per cent in Bangladesh to only 35 per cent in Sri Lanka. However, official statistics often grossly underestimate the female workforce in the region. Women's work, in the household, the farm or the commons is labelled as household chores and not given the status of work. With more expanded definitions of economic activities being introduced recently, it has been observed that the 'statistical invisibility' of women workers in South Asia is reducing. For example, in Pakistan, with the adoption of the UN definition of the system of national accounts (SNA) boundary, women's participation in the workforce has increased from 13.7 to 39.2 per cent. In Bangladesh, the corresponding figures are from 18.1 to 50.6 per cent (1996) (HDSA 2003).

Independent of definitional issues, the sectoral distribution pattern of the female workforce reveals that the agricultural sector employs over two-thirds of the women workers in all South Asia countries with the exception of Sri Lanka (49 percent) and the Maldives (28 per cent during 1980-90; see Table 4). In addition, what makes women's contribution to macro level food security even more significant is the fact that in most countries, women, by choice or restriction, focus largely on subsistence production of food crops, be it on farms or in home gardens, whereas their male counterparts tend to diversify into commercial farming.

Given women's role in food production and provision, any set of strategies for sustainable food security must address their limited access to productive resources. Women's limited access to resources and their insufficient purchasing power are products of a series of inter-related social, economic and cultural factors that force them into a subordinate role to the detriment of their own development and that of society as a whole (FAO, 1996)

A serious constraint for women farmers is their lack of access to security of tenure or ownership of land. As Agarwal (2002a) notes, 'In largely agrarian economies, arable land is the most valued form of property and productive resource. It is a wealth creating and livelihood-sustaining asset. For a significant majority of rural households, it is the single most important source of security against poverty'. South Asia falls in the male farming system category and is part of the belt of classic patriarchy characterized by extreme forms of gender discrimination (IDRC 2004). This includes the right to ownership of land.

Traditionally, women have been denied equal inheritance rights to property both under the Hindu and the Islamic systems of law. In India, for example, under the Hindu system, a woman could inherit property only in the absence of four generations of men in the male line of descent. Even then, her rights were limited to a lifetime interest with no right to mortgage or dispose of the property, except in exceptional circumstances. The Hindu Succession Act of 1956 sought to resolve this issue by giving equal shares to sons, daughters and widows in a man's own property and the same in his share of joint family property. However, agricultural land was exempted from the purview of the Act. Women's inheritance in agricultural tenancy land depends on state-level tenurial laws, which strongly favour succession in the male line.

Under Muslim law, daughters are allowed only half the share of sons in the property bequeathed by their father. The Muslim Personal Law Shariat (Application) Act of 1937 also sought to enhance the property rights of Muslim women, but excluded all agricultural land, both tenanted and owned from its purview, except in some of the states of South India.

Discriminatory patterns of land ownership extend right across the South Asian region. Pakistan and North West India are characterized by the severest gender-based inequities, which tend to reduce towards the southern and north-eastern parts of the region. These variations are, in part, due to pockets of matrilineal systems prevailing in the southern state of Kerala and the north-eastern state of Meghalaya, as well in the kingdom of Bhutan where land is inherited in the female line, and in part due to the underlying social constructs in these regions. Social norms in the southern part of India permit marriage within kinship and in the same village, where in the northern part of India, marriage within the family is strictly prohibited with intra-village marriage relationships being frowned upon. The impact of these social strictures can be clearly seen in the strictly patrilineal inheritance patterns in the north as compared to the more flexible rules in the south. In fact, in the southern states of Andhra Pradesh, Tamil Nadu and Karnataka, as well as in Maharashtra in central India, the Hindu Succession Act has been amended to make daughters coparceners in joint family property, while in the southern state of Kerala, the concept of joint property has been abolished altogether.

Unlike other countries in the region, Bhutan allows no overt discrimination against women socially, economically, politically or legally. Women are accorded a dominant role in the legal system especially in family and inheritance law. The law of inheritance reserves equal rights for all children and in many parts of Bhutan, society is matrilineal with women inheriting land (UNCT, 2000).

The gender discriminatory pattern of inheritance is reiterated in the remaining countries of the region, i.e, Nepal, Bangladesh and Sri Lanka. While in all these countries, there is no constitutional barrier to gender equality, personal laws govern inheritance, marriage and other social contracts and are in most cases inherently discriminatory. In Bangladesh, inheritance rights are governed by religion and under all religious laws, women have a lesser share than men. Under Muslim law, a daughter inherits one-half the share of her brother, a wife can claim one-eighth of the property, while a mother gets one-sixth. Under Hindu laws, as in India, women enjoy their right to property only for their lifetime. It cannot be transferred or inherited by daughters and reverts back to family ownership after the death of the women. Under Christianity, the situation is slightly better. However, it is common in rural areas for daughters to renounce their claim to family land or accept a lump sum payment in lieu of their property rights, merely to preserve visitation rights to the parental home. This practice is also common in northern India, where women often turn to their brothers for support in case of widowhood or desertion by their husbands. As women are not perceived as having an existence independent of a man, their rights to land, housing or other assets is not recognized (Khatun, ed.).

In Sri Lanka, several parallel systems of personal laws based on differing social and cultural practices of ethnic and religious groups coexist, which deny women equal status with men in respect to property. These include the General Law, the Kandyan Law, the Thesawalamai and the Muslim Law. These laws were codified during colonial times and successive post-independence governments have guaranteed their continued existence (Chulani, 3003). Under the

Thesawalamai system of personal law, which is applicable to the Malabar inhabitants of Jaffna, a married woman cannot dispose of her immovable property without the consent of her husband. While there has been considerable reform of the General Law to ensure equality for women, the Muslim Law continues to show a preference for males intestate succession.

In Nepal, women traditionally have exclusive rights to two types of property: *daijo* or the small plots of land and other immovable property that are sometimes given to them on marriage and *pewa* or anything given to a woman as her personal property or that she earns herself. In practice, however, there is a frequent lack of land titles in women's names (Trencher & Shreshta, 2002). A wife cannot inherit land without the consent of her husband or son. In the case of unmarried daughters, the consent of the father is needed (CEDAW 2004).

Gender equality in inheritance, Agarwal (2002a) writes, must be promoted as most agricultural land is privately held. In India, for example, 86 per cent of the arable land is in private hands. While gender disaggregated records are not maintained, sample surveys have brought to light the sharply skewed pattern of inheritance in rural areas of India. Chen's 1991 (Chen, 2000) survey of seven states in India, quoted in Agarwal, found that of 470 women with land owning fathers, only 13 per cent inherited land as daughters. This ranged from 18 per cent in south India to only 8 per cent in the north, indicating that 87 per cent of women did not receive their due as daughters. Of the 280 widows surveyed, only 51 per cent received land and most often their shares were not recorded in official land records. Other studies reveal that even when women's land rights are recorded, it is usually in joint ownership with their sons.

Land reform programmes and resettlement schemes are also overtly male biased. Under 'operation Barga', a scheme implemented by the West Bengal state government to secure the rights of tenants by registration in the late 1970s and early 1980s, few women received land. A study conducted in a village in Midnapur district (Gupta, 1993 quoted in Agarwal (2002a) showed that 98 per cent of the 107 holdings distributed, went to men and in 9 out of 10 female-headed households, it went to the sons. Married women did not even receive joint titles. Under the eighth five year plan (1992-97), state governments were directed to allot 40 per cent of ceiling surplus land to women alone and the rest as joint titles to spouses. However, ceiling surplus land accounted for just about one-half per cent of India's arable land in the early 1990s and today is reduced to merely 0.2 per cent. Also, the implementation of the scheme rests with state government, which may or may not follow through.

Why is ownership of land so essential for women farmers? The rapid feminization of agriculture in the region has thrown into prominence the issue of land rights for women. Increasing migration by males from rural to urban areas in search of livelihoods has followed the fragmentation of land holdings, lack of wage opportunities in rural areas and deepening poverty. What is often overlooked in policy formulation is the increasing number of de facto woman heads of households struggling to eke out a livelihood and ensure the food security of their families without access to credit, technology or extension services. Denied security of tenure, they lack the collateral required for credit or the social status to deal with extension workers on an equal basis. Their needs tend to be ignored, even in agricultural research and technological innovations. FAO statistics show that worldwide, only 5 per cent of extension services have been addressed to rural women (FAO 1996).

In order for women farmers to use land more efficiently and thereby make a greater contribution to food security, they need access to land, management control of land-based resources and the economic incentives that the security of tenure provides (FAO 1996). Land rights can serve multiple functions in rural women's lives, which are not easy to replicate through other means. Apart from the direct benefits in terms of crop output, trees, fodder, fuel and garden produce and indirect advantages in terms of collateral for credit or an asset, which can be sold or mortgaged when needed, title to land also enhances the probability of finding supplementary wage employment and serves as an asset base for rural non-farm enterprises. Chadha (1992), in a study of the rural non-farm sector, finds that those with land generated much higher rural non-farm earnings from self-employment than the totally landless. Agarwal concludes that, women's access to even a small plot can be a critical element in a diversified livelihood system and can significantly improve women's and the family's welfare, even if the plot is not large enough to provide full family subsistence. An IFPRI study (Meizen-Dick, 2004) finds that women with land in

Bangladesh were affered higher wages for working on other fields. An IFAD study of Bangladesh in 2000 (cited in Agarwal, 2002a) identified lack of access to land and homesteads as the major factors in exclusion of the poor from credit NGOs.

LIMITED LIVELIHOOD OPTIONS

Female labour force participation tends to be associated with poorer households in most of the South Asian region, as rural women from better-off households tend to work in home-based activities which are less likely to be picked up by labour force statistics (Caber, 2003). It has been observed that while poverty may force women to work outside the home, increased household prosperity may lead to their withdrawal from the labour force once again. At the same time, the active role that women, including those from better-of households, play in home-based economic activities tends to be socially and statistically invisible, accounting for the extremely low labour forced participation rates of women throughout South Asia and also to the strong relationship between paid work by women and household poverty (IDRC 2004). In the region as a whole, women account for about 33.5 per cent of the workforce as compared to 42 per cent in SSA and 44.5 per cent in East Asia and the Pacific (1995-2001). In terms of female economic activity rates, the contrast with other developing regions is even sharper, with South Asia recording only 42.7 per cent as against 62.3 and 68.9 per cent in Sub-Saharan Africa and East Asia and the pacific, respectively (HDSA 2003).

The employment status of women workers in the region could provide a clue as to their unequal economic and social status. With the exception of Sri Lanka, the largest segment of the female workforce is employed as contributing family workers, i.e., engaged in unpaid and often unacknowledged economic activity. This segment comprises as much as half to three-quarters of the women workers. By far, the worst off in this regard is Bangladesh, where as many as 77 per cent of the female workers are employed in family-based economic activity. Waged and salaried workers (with access to cash incomes) constituting the more empowered sections of the workforce, account for as little as 7-8 per cent of the female workers in Nepal and Bangladesh and about one-third in Pakistan. It is only in Sri Lanka, where women more than equal their male counterparts in wage-earning and salaried activities, with 68 per cent of the female labour force employed in this sector as compared to only 60 per cent of the men. Where self-employment is concerned, a substantive proportion of the female workforce in Nepal, i.e., 29 percent fall in this category, while in the other countries, the proportion of self-employed women ranges from 8 to 16 per cent.

The largest segment of the female workforce throughout South Asia is employed in the agricultural sector. In countries like Bhutan and Nepal, almost all women workers (98 per cent and 90 per cent, respectively) are engaged in agriculture. The Maldives is the only exception with over half its female workers employed in the industrial sector and only about one-third in agriculture. In the remaining countries of the region, the service sector employs a larger proportion of the female workforce than the industrial sector. However, in India, the employment share of the industrial sector is roughly equivalent to that of the service sector (see Table 4).

In 1999-2000, the share of the informal sector in total non-agricultural employment in South Asia was the highest in the world. Compared to the 1980s, this share went up by 55 per cent, with 55 per cent women and 48 per cent men, self-employed. The reason for this sharp increase in self-employment is generally attributed to the structural adjustment programmes of the 1990s that led to the reduction of public investment, the cutting back of public sector jobs and the increase in the demand for subcontracted flexible labour to produce goods for local markets. Informal wage employment is estimated to account for 30-40 per cent of informal employment in the non-agricultural sector. This includes casual day labourers, part-time or temporary workers without contracts or social security, domestic workers, industrial outworkers, etc. India and Bangladesh are the only South Asian countries, which collect data on casual workers. Both these countries show an increasing trend of female casual workers (Human Development in South Asia 2003). The growing actualization of female workers in India-also referred to as the feminization of poverty-accompanied by the increasing gap in wages between men and women, as well as the duration of work available in terms of days, has a significant [negative] impact on household food security (Raju, 2001).

Table 4. Female employment by sectors 1991-99 in South Asian Countries

Country	Agriculture	Industry	Services
Bangladesh	77.5	7.6	11.0
Bhutan *	98.1	0.3	1.5
India	77.7	11.1	11.3
Pakistan	66.3	10.5	23.1
Maldives *	28.2	51.3	20.5
Nepal	90.5	1.3	8.2
Sri Lanka	48.8	22.2	27.3

Note: * Data for Bhutan and the Maldives refer to the decade 1980-90.
Source: HDSA (2003).

WAGE DIFFERENTIALS

Throughout South Asia, women's wages range from half to two-thirds of wages received by men. This inequity in the wage structure is particularly marked in Pakistan where women receive, on an average, just about one-third of the wages paid to men, and in India where the wage differential is as high as 38 per cent (Table 5). The Maldives and Nepal show the least disparity in the wage structure, but nowhere in the region do women receive more than 60 per cent of the wages paid to men.

Table 5. Estimated wage differentials between women and men (PPP US$) in 2001

Country	Female	Male	Wage differential in %
Bangladesh	1,153	2,044	56.41
India	1,531	4,070	37.62
Maldives	3,329	5,582	59.64
Nepal	887	1,734	59.64
Pakistan	909	2,824	32.19
Sri Lanka	2,295	4,189	50.01

Source : UNDP (1998,2002,2003)

In the case of India, gender-specific wage rates for both agricultural and non-agricultural operations averaged over 600 sample villages across 20 states, (GOI, 2001) have been compared by Ramachandran (2003). As wage rates fluctuate with the seasons, for purposes of comparison only the maximum wage rate for each activity during the agricultural year has been tabulated (Table 6). It is evident from the table that the wage rates paid to women workers in the agricultural sector are at least 20 to 30 per cent lower than those paid to men for the same activity. In non-agricultural activities, the difference is even more pronounced, with women being paid less than half the wages given to their male counterparts. Surprisingly, even in the case of activities like cotton picking and tea leaf plucking, where women undoubtedly have the edge and female labour is preferred to male, the disparity in wages persists, though the difference is less marked.

Micro studies undertaken in various states of India reveal conflicting findings. Surveys undertaken in West Bengal (cited in IDRC 2004) show that gender wage disparities have begun to decline along with the general rise in wages between the mid 1960s and early 1970s. Female wages in the state rose from 75 per cent of male wages to about 86 per cent. The study states that on an average, across India, rural women's wages rose from 52 per cent of their male counterparts in 1972 to as much as 69 per cent by 1983. The gap between male and female wages was highest and fluctuated most in the least agriculturally developed areas. However, a more recent micro-study of 54 casual labour households in the Karimnagar district of Andhra Pradesh (Lingamurthy, 2002) finds that while both male and female wages have increased over time, the relative wage differentials by gender have not reduced. Females are paid less than half the wages given to men across all agricultural and non-agricultural operations. At best, women receive one-half of the male wages for the same operation.

Table 6. Gender specific wage rates for agricultural and non-agricultural occupations in India (1999-2000)

Activities	Maximum daily wage (in INR)	
	Men	Women
Ploughing	72.08	49.25
Sowing	66.64	52.85
Weeding	56.94	49.56
Transplanting	58.33	50.50
Harvesting	60.16	51.35
Winnowing	54.51	45.11
Threshing	60.55	47.72
Tea plucking, cotton picking, etc.	49.47	40.85
Herding	43.15	34.41
Well-digging	76.18	45.23
Cane crushing	56.67	36.04
Blacksmithy	85.41	40.0
Masonry	110.10	43.03

Source: Government of India (2000/1).

In Sri Lanka, while the disparity between male ad female wages for the same activity is less marked overall, there are considerable inter-regional variations in wage structure. The rapid growth of the agricultural sector in Anuradhapura and Polonnaruwa has increased female participation in farming, especially, in paddy and non-traditional crop. As a consequence, women's earnings from agriculture are high-as much as 89 per cent of male earnings in Anuradhapura and 87 per cent in Polonnaruwa. In Nuwara Eliya, where women are employed on tea plantations, minimum wage regulations are enforced and periodically revised and women earn about 89 per cent of the male wage. In other districts like Puttalai, Kegalle and Moneragala, however, the wage disparity is more in line with the rest of South Asia, with women receiving less than 50 per cent of male wages (UNDP 1998).

In the case of Pakistan, a sample survey of rural Punjab cited in the PNHDR (2003) reveals that among farm households, only 35 per cent of the women in the labour force are engaged in paid work and even these receive meagre returns as they are relegated to low-paid farm labour. The study also shows that while women work on an average for 101 days a year, their daily income averages at US$ 15, far below subsistence level.

WOMEN 'S CONTRIBUTION TO HOUSEHOLD INCOME AND ITS IMPACT ON FOOD SECURITY AND NUTRITION

Women from poor households engage in variety of income-generating and expenditure-saving activities. In some cases these supplement male contributions, while in others they are the primary or sole source of household livelihoods. Women's work is often critical to the survival and security of poor households. Despite the low pay that they receive, their contributions have been shown to be the single most important element in the survival strategy of both rural and urban households. A study (National Institute of Urban Affairs 1998 cited in IDRC 2004) covering six cities in India confirms that 11 per cent of poor households relied entirely on female earnings, while women contributed 25-50 per cent of the earnings of about one-third of the households. In Bangladesh, wage-earning women from landless households contribute about 24 per cent of the household's total annual income (IDRC 2004). In rural Pakistan, also, women from poor households have been found to contribute between 16 to 25 per cent of the family income (PNHDR 2003).

Households, wherein women have access to their own incomes and can exercise decision-making powers, tend to have an expenditure pattern different to the one existing in male dominated households. Research in several developing countries of Asia, Africa and Latin America has found that improvements in household food security and nutrition are associated with women's access to income and their role in household decisions on expenditure. This is because women tend to spend a significantly higher proportion of their income than men on food for the family (IDRC 2004). Dyer and Bruce (cited in Agarwal, 2002a) in a 1988 study find that women in poor households of India and other parts of the world, spend most of the earnings under their control on basic household needs, while men tend to spend a significant part of theirs on personal good such as alcohol, tobacco, etc. Direct evidence from South Asia, though limited, presents a similar picture. Studies undertaken in India and neighbouring countries reveal that women invest larger amounts of money in nutrition and health. Among marginal farmer households in the Indian state of Kerala, the mother's cultivation of a home garden (the output of which she controlled) is found to have a consistently high positive effect on child nutrition (Kumar, 1978).

A recent study carried out in Bangladesh states that in the use of income from credit-based structures, women mentioned a greater expenditure for food, clothes, children's education and health (Kelkar, Nathan & Jahan, 2003). In rural Pakistan, the majority of women borrowers (94 per cent) borrowed to fulfil the consumption needs of the household (PIDE 1998 cited in Nazli & Hahmed nd). This has led many policy makers and donors to conclude that women should be targeted for credit and small enterprise programmes not only because their income boosts houshold income, but also because it meets global societal objectives such as increased spending on food and children's goods (Hopkins *et al.*, 1994 cited in Ramachandran 2004).

INTRA-HOUSEHOLD GENDER DISPARITIES AND INDIVIDUAL ACCESS TO FOOD

It is now widely accepted that 'gender inequality dwells not only outside the household, but centrally within it' (Agarwal 2002b). Mainstream economic theory that accepted the household as a unitary entity, wherein resources and incomes are pooled and household members share common interests and preference has been effectively challenged on the basis of recent empirical findings. As Agarwal (2002b: 2) states, 'Gender, in particular, is noted to be an important signifier of differences in interests and preferences, incomes are not necessarily pooled and self-interest resides as much within the home as in the market place, with bargaining power affecting the allocation of who gets what and who does what'.

Not only do intra-household power equations serve to keep women unempowered and subservient, but also directly impact on their individual food and nutrition security and indirectly on that of other family members, particularly children. Within the context of household dynamics, food security is related to decisions regarding responsibility for food production, earning cash income or food purchasing and preparing food and finally, actual access to food in terms of consumption.

It is often difficult to assess the gender disparity in access to food within the household as differences in calorie consumption (the standard method of accessing food intake) may be attributed to the lower energy needs of women. However, indirect evidence in terms of gender-specific malnutrition levels point to existing disparities. In poor households, in particular, the incidence of severe malnutrition is greater among girls. In fact, gender has been found to be the most statistically significant determinant of malnutrition among young children, and the most common cause of death among girls below the age of five years. Studies from the state of Punjab in India have shown a sharp difference in calorie intake among adult men and women, with women consuming approximately 1000 fewer calories than men (Development Gateway 2004). A study of eleven villages in Punjab (Dasgupta, 1987 quoted in Bose 2003) finds that though boys and girls had roughly similar calorie intake, girls were given more cereals, while boys were given more milk and fats with their cereal. The study also observes that discrimination against the girl child was primarily motivated not by economic hardship, but rather by cultural factors.

Intra-household gender bias in favour of male children, both in terms of feeding and seeking healthcare has been noted in Pakistan (Nazli & Hamid, nd). Kabber (1998) (cited in Rahman, 2002) finds that women in Bangladesh are quite literally a residual category in intra-household food distribution, eating after men and the children, and making do with what is left. This deprivation is partly self-imposed and is handed on from generation to generation. A similar pattern prevails in most South Asian countries. A study of tribal villages in four states of India (Barme & Ramachandran, 2002) notes that coping strategies adopted by households faced with seasonal food shortages involved reduced food consumption by women as a first step followed by skipping of meals in order to ensure that the male members of the family and the children had larger portions of food from the meagre store (Ramachandran, 2004). Even pregnant women are caught up in the cycle of self-denial and food deprivation. A study of 177 women in various stages of pregnancy in rural West Bengal (Mondal, 2003) shows that the structure of the family also plays a role in female nutrition. In nuclear families where the women herself has the responsibility of food distribution, she gives preference to her husband and children at the cost of her own needs. It is only in joint families, where a mother-in-law is present that the nutritional needs of a pregnant women in terms of access to more nutritious food, are better taken care of, even if she does eat after the earning members.

Rahman (2002) goes beyond the calorie trap and analyses data on the quality of food consumed in the household. Based on data from a household survey by IFPRI (1996-97) of 47 villages in Bangladesh, he notes that while pre-school children are the most privileged family members in terms of expensive energy food such as meat, fish and dairy products, gender differences are perceptible even at this stage, with boys being favoured over girls. What is even more interesting is the fact that the wife's assets brought as dowry at the time of marriage seem to influence her bargaining power within the household, as well as her access to better food. Rahman finds that an increase of 1000 *taka* increases the adult women's food security/energy security index for animal, dairy and fish by about 25 per cent. Another significant finding of the study is the fact that while gender disparities among pre-school children tend to disappear in the middle- and higher-income groups, the neglect of the adult female persists across all groups.

DIFFERENTIAL ACCESS TO NUTRITION AND ITS INTER-GENERATIONAL CONSEQUENCES

Women's access to adequate food security, both for themselves and their families, is dependent not only on their economic status, but on their own health, education and social status within the family and in society. Women with low status tend to have weaker control over household resources, tighter time constraints, less access to information and health services, poorer mental health and lower self-esteem. These factors are thought to be closely tied to woman's own nutritional status and the quality of care they receive (Smith *et al.*, 2003). Smith concludes that making a decision at the policy level to improve women's own nutritional status produces significant benefits. Not only does a women's nutritional status improve, but so does the nutritional status of her young children. Raising women's status today is a powerful force for improving the health, longevity, mental and physical capacity and productivity of the next generation of adults (Smith *et al.*, 2003).

Overall, the South Asian region is characterized by high fertility rates, negative sex ratios and unacceptably high levels of maternal undernutrition and mortality. The exception to the general trend is Sri Lanka which, as pointed out earlier, has consistently been cited as a model for the rest of the region. Among the remaining six countries, fertility rates have reduced in India and in Bangladesh, but they are still about one and a half times higher than those of Sri Lanka. The other countries of the region including Bhutan, Pakistan and the Maldives still record fertility rates of 5 and above, maternal mortality rates (MMR), which accurately mirror the level of under nutrition and lack of access to healthcare facilities are high throughout the region, but particularly so in India, Nepal and Bangladesh. Bhutan's enormously high MMR of 1600 could be attributed partly to the unavailability of current data, but largely to the lack of amenities and the remote and inaccessible locations of many settlements as well as the fact that 90 per cent of deliveries are home based.

Nutritional deprivation has two major consequences for women– they never reach their full growth potential and they suffer from anaemia. Both are risk factors in pregnancy. High levels of anaemia complicate childbearing and result in maternal and infant deaths and low birth weight infants (Chatterjee, 1990 cited in Coonrod 1998). The magnitude of the problem can be gauged by the fact that about half of all anaemic women in the world live in this region. India, with 88 per cent of all pregnant women developing iron deficiency anaemia, has the worst-case scenario in this regard. Anaemia impairs human functions at all stages of the life span. Severe anaemia during pregnancy increases the risk of maternal mortality by up to 23 per cent in Asia (Ross & Thomas cited in Gillespie & Haddad, 2003). In areas marked by high under-nutrition, malnourished women or adolescent girls give birth to babies that are born stunted and thin. In this way, undernutrition is handed down from one generation to another as a terrifying inheritance (Gillespie & Haddad, 2003).

An additional factor affecting women's nutritional status is the seasonal dimension reflecting fluctuations in their workloads during the agricultural cycle. The period of greatest nutritional stress for rural women is the 'lean months' of the pre-harvest period when household stocks and energy intake are low, but the energy demands of agricultural work tend to be highest. Heavy work during pregnancy can lead to premature labour and without increased calorfic intake, to low birth weight babies (IDRC 2004). A participatory study of tribal villages in four states of India (Barme & Ramachandran, 2002) reiterates these findings and concludes that the impact of several months of reduced food intake on the health and nutritional status of women, particularly pregnant and lactating mothers and on their new born infants is reflected in the persistently high levels of low birth weight infants, malnourished and anaemic mothers and maternal mortality in the country (Ramachandran, 2004).

Over one-quarter to half of the infants in the region are born underweight, i.e., with a birth weight of less than 2.5 kg. Even Sri Lanka, which has consistently better human development indicators than the other countries in the region, records one fourth of all newborn infant as having less than the minimum weight. In this regard, the Maldives rivals other more developed countries with only 13 per cent of its infants being born with low weights. In the case of child malnourishment, all the countries in the region record over one-third of children below five years as undernourished. As with low birth weights, Bangladesh and India fare worst, with over half of all children below five years of age being undernourished, closely followed by Nepal. The prevalence of low birth weight is strongly associated with the relative under-nutrition of mothers in the region. It is estimated that about 60 per cent of the women in South Asia are underweight, i.e., they weigh less than 45 kg. Early childhood growth failure is manifested by growth stunting, which is a risk factor for increased mortality, poor cognitive and motor development and other impairments in function. Children who have been severely undernourished in early childhood suffer a later reduction in IQ by as many as 15 points (Marmoreal, 1996 cited in Gillespie & Haddad, 2003), significantly affecting schooling achievement and increasing the risks of dropout or repeat grades. Moreover, stunting usually persists, leading to smaller size and poorer performance in adulthood. While the most common cause of poor growth is poor maternal nutrition status at conception and *in-utero* under nutrition, other contributing factors include inadequate breast-feeding, delayed complimentary feeding for infants, impaired absorption due to infections or a combination of these problems. Underpinning these factors are various inadequacies with respect to household and community level access to food, health, environmental and caring resources.

Age-specific comparisons of male-female mortality show that the disadvantage suffered by South Asia women is not a simple biological phenomenon that begins at birth (Osmani, 1997). As Osmani points out, when under-5 mortality rates are broken up into neo-natal mortality (first seven days of life), post-natal mortality (seven days to one year), infant mortality (up to one year) and child mortality (between one and five years), it is evident that gender discrepancies emerge later. For example, in India, the post-natal mortality rate is 36 per 1000 for females as against 32 for males –a rather small difference which, however, increases to 42 for females as against 29 for males when child mortality is considered. He concludes that the origin of female disadvantage lies in the discriminatory treatment meted out to women in the allocation of life saving resources such as food and healthcare. A recent World Bank Report on public policy and Service Delivery in India (World Bank 2004) reinforce these findings and states that parental neglect towards girls, symptomatic of the generally low social status of women, appears to be an important cause of the gender disparity in child mortality. The study finds that there are sharp gender disparities not only in terms of medical treatment but also when availing of free nutritional inputs provided in ICDS centres. An analysis of NFHS I data showed that, after controlling for other factors, all positive nutritional benefits seem to have accrued to boys, suggesting that parents selectively bring only boys to the ICDS centre. Healthcare and morbidity and medical treatment records show that female babies in South Asia are less likely to be vaccinated or treated for acute respiratory infection (ARI) and fever. Osmani's study further states that more direct evidence comes from gender differences in the number of effective life years lost due to premature death and disability from illness. Almost everywhere in the region women in the reproductive age group (15-44 years) suffer more than men, but in India the disparity is highest with the female-male ratio of effective life years lost due to illness related disability being as high as 1.6 as against 1.3 for the other developing countries overall.

Female literacy is now widely recognized to be an important determinant of the health of a nation. An analysis of countrywide household surveys in India, Pakistan and Sri Lanka tabulated the incidence of child under nutrition separately for four categories of mother's education. It was evident that in each country, for all three measures of child under-nutrition, the under nutrition falls monotonically with the level of maternal education: the illiterate mothers being associated with the highest incidence in every case. Even women who have not gone beyond primary school can have as much as 20 per cent less under-nutrition among their children as compared to illiterate mothers (Osmani, 1997). Data from 25 developing countries suggests that 1-3 years of maternal schooling reduces child mortality by 15 per cent, while an equivalent paternal schooling achieves only a 6 per cent reduction (IDRC 2004). Lennart Bage, president, IFAD states:

"Issues of caring practices are of growing importance, as these are typically in the domain of women. This is not the only link between gender, household food security and nutrition, but is a particularly strong one. Investing in the education of women through primary schooling for girls, functional literacy for adult women or nutrition/ health education for women's groups generates multiple positive effects. Caring practices tend to improve, as do most indicators of family wellbeing (Bage, 2000)"

A study exploring the relationship between women's status and children's nutrition in three developing regions-South Asia, sub-Saharan Africa and Latin America and the Caribbean (Smith *et al.*, 2003) notes that increases in women's status in South Asia have a strong influence on both the long-and short-term nutrition status of children leading to reductions in both stunting and wasting. The human costs of women's lower status in the region are high. The study estimates that if women and men had equal status, the child underweight rate would drop by approximately 13 percentage points, meaning 13.4 million fewer malnourished children in this age group alone.

Throughout the world, it has been shown that improving healthcare for women aged 15-44 years offers the biggest returns on healthcare spending of any group of adults (ADB 2004).

CONCLUSIONS

The foregoing analysis provides sufficient evidence to indicate that the role of women in ensuring food security at macro level as well as at the level of the household and the individual is a manifold one. It is also apparent that in South Asia, at every stage of this relationship, deeply embedded social constructs impact adversely on woman's economic contribution to society as well as her nutrition and health status, and by extension, that of her family and society at large. While much progress has been made on the food production and availability front, adequate nutrition outcomes cannot be assured without unraveling the complexities of the gender food security link. Ensuring equity in women's rights to land, property, capital assets, wages and livelihood opportunities would undoubtedly impact positively on the issue, but underlying the deep inequity in woman's access to nutrition is her own unquestioning acceptance of her status as an unequal member of the family and society. Eventually, gender empowerment alone is likely to be the key to the resolution of the hunger challenge in the region.

While women's contribution to economic output is slowly being recognized and in many cases women have greater access to education and livelihood opportunities, changes in economy and society are throwing up new challenges to ensuring freedom from hunger and malnutrition. The sharp increase in women-headed households has brought into focus the issue of land rights for women as also the right to housing and other assets. In short, the emerging issues may be summarized as follows:

- The issue of women's entitlement to productive resources including land, livestock, agricultural implements and credit remains a major obstacle in the productivity of female and to a large extent, the maintenance and sustainability of cultivable land;
- The South Asian experience reviewed above reveals that gender wage disparities are sharpest in the least developed agricultural areas and tend to decrease with agricultural growth and increased employment opportunities, as evidenced in the case of West Bengal and Sri Lanka. With the growing incidence of women-headed households, the fact that a large proportion of households below the poverty line are headed by women and the substantial contribution to household income is made by women from the poorest households, the issue remains a major one in the endeavour to ensure household food security;
- Gender differentiated expenditure patterns from a selection of developing countries further emphasize the importance of women's access to and control over household income in efforts to enhance household food and nutrition levels;
- Education has emerged as the single most important factor affecting individual food and nutrition security. The strong positive correlation between literacy and various education levels of mothers with children's nutrition levels has been verified in various studies across diverse developing countries and has also been shown to have the largest impact in the South Asian region;
- Ultimately, gender inequities in food and nutrition security lie at the root of the cycle of hunger and malnutrition in the region. It is, in the ultimate analysis, an issue of woman's own perception of her status. Only when women in the region begin to feel

empowered and equal in status to men, will the stranglehold of gender disparities across the region weaken and break. It is then that food security will become merely an economic issue with simple solutions to the problem.

References

Agarwal, B. 2002a. Are We Not Peasants Too? Land Rights and Women's Claims in India. *Seeds*. 21: 29. Available at www.popcouncil.org/publications/seeds/seds21.pdf.

Agarwal, B. 2002b. *'Gender, Inequality, Cooperation and Environmental Sustainability'*. Available at: www.santafe.edu/research/publications/workingpapers/02-10-058.pdf.

Asian Development Bank (ADB) 2004. Gender and Development Issues in the Asia and Pacific Region. In : *Policy Gender and Development: Policies and Strategies*. Available at: www.adb.org/documents/policies/gender0200.asp.

Bage, L. 2000. The Well Fed has Many Problems, The Hungry only One. *United Nations Chronicle*. 36 (3). Rome: IFAD, Department of Public Information.

Barme, C. & N. Ramachandran. 2002. *Foodgrain Bank Project: Report of the PRA Conducted in Selected States and Final Recommendations*. Rome: WEP-FAO.

Bose, A. B. 2003. *The State of Children in India: Promises to Keep:* Manohar, New Delhi.

Cedaw. 2004. *Press Release-Nepal*. 630 and 631st Meetings. Available at: www.unorg/News/Press/docs/2004/wom1423.doc.htm.

Chadha, G. K. 1992. *Non-Farm Sector in India's Rural Economy: policy, performance and Growth prospects*. Centre for Regional Development, Jawaharalal Nehru University, Delhi. Mimeo.

Coodnrod, C. S. 1998. Chronic Hunger and the Status of Women in India. Available at:www. Thp. Org/reports/indiawom. Htm.

Chen, M. 2000. *Perpetual Mourning: widowhood in Rural India*: University of Pennsylvania press, Philadelphia

Chulani, K. 2003. Engaging with Muslim personal Law in Sri Lanka: The Experience of MWRAF. *Lines*. August Available at: www.lines-magazine.org/Art_Aug03/Chulani.htm.

Development Gateway 2004. News on Food Security, Intra Household Gender Disparities and Access to Food. Available at www//topics.developmentgateway.org/gender.

FAO 1996. FAO Focus: Women and Food Security: Women Hold the Key to Food Security Available at www.fao.org/FOCUS/E/Women/WoHm-e.htm.

Gillespie, S. & L. J. Haddad. 2003. *The Double Burden of Malnutrition in Asia–Causes, Consequences and Solutions*. Sage Publications, Newbury Park.

Gupta, J. 1993. Land, Dowry, Labour: Women in the Changing Economy of Midnapur. *Social Scientist* 21 (9-11): 74-90.

HDSA 2002. *Human Development in South Asia 2002: Profile of Agriculture in South Asia* : Oxford University Press for Mahbub ul Haq Human Development Centre, Islamabad.

HDSA 2003 *Human Development in South Asia 2003*. Oxford University Press for Mahbub ul Haq Human Development Centre, Islamabad.

IDRC 2004. *Gender Inequality and Poverty Eradication*: Promoting Household Livelihoods. Available at: *web.idrc.ca/fr/er-42966-201-1-DO_topic.html.*

Government of India (GOl) 2000/1. *Indian Labour Yearbook 2000-01*: All-India Average Daily Wage Rates for Agricultural and Non-Agricultural Occupations for the Agricultural Year 1999-2000 (July-June). Government of India, New Delhi.

Kabeer, N. 1998. Subordination and Struggle: Women in Bangladesh. *New Left Review* 168: 95-121.

Kabeer, N. 2003. Safety Nets and Safety Ropes: Addressing Vulnerability and Enhancing Productivity in South Asia. In : S. Cook, N. Kabeer & G. Suwannarat (Eds.) *Social Protection in Asia.: The Ford Foundation*. Har Anand Publications Pvt. Ltd., New York and New Delhi.

Kelkar, G., D. Nathan & R. Jahan. 2003. *We were in Fire, Now We Are in Water: Micro-Credit and Gender Relations in Rural Bangladesh*. Working Paper Series No. 19. Institute for Human Development, New Delhi.

Khatun, S. M. (Ed.) Women and Shelter Development: Perspectives and Reality in Bangladesh Context Paper presented at the Seminar on United Nations and Women. Dhaka.

Kumar, S.K. 1978. *Role of the Household Economy in Child Nutrition at Low Income*. Occasional Paper No.95. Department of Agricultural Economics. Cornell University, Ithaca

Lingamurthy, S. V. N. 2002. Gender Inequalities in the Wages of Agricultural Labourers: A Micro Study. *The Asian Economic Review*. 444 (2): 333-38.

Meizen-dick, R. 2004. Omen's Property Rights and Food Security. Washington, DC:IFPRI. Available at www.developmentgateway.org/download/237035/ruth.pdf.

Mondal, S.K. 2003. *Health, Nutrition and Morbidity, a Study of Maternal Behaviour*. Bookwell, New Delhi.

Nazli, H. & S. Hamid (Ed). *Concerns of Food Security, Role of Gender and Intra Household Dynamics in Pakistan*. Pakistan Institute of Development Economics (PIDE), Islamabad. Available at: www.pide.org.pk/research/report175.pdf.

Osmani, S.R. 1997. *Poverty and Nutrition in South Asia*. The First Dr. Abraham Horwitz Lecture, United Nations Standing Committee on Nutrition. Available at: www.unsystem.org/scn/publications/horwitzlecture/lecturenol.htm.

PNHDR (Pakistan: National Human Development Report) 2003. *Poverty, Growth and Governance*. UNDP, Karachi.

Rahman, A. 2002. *On Measuring Intra household Inequality in Food Distribution – Is the Conventional Calorie Intake Enough to Understand Individual Well being within the Household?*. Department of Economics, University College, London. Mimeo.

Raju, S. 2001. Sustainable Livelihood, Gender and Food Secutity In :M.D. Asthana and P. Medrano (Eds.): *Towards Hunger Free India: Agenda and Imperatives*. Manohar Publishers, New Delhi.

Ramachandran, N. 2003. The Paradox of Gender Food Security. Paper presented at the Gender Poverty Summit, Women's Political Watch and National Council for Applied Economic Research, 9-11 November. New Delhi.

Ramachandran, N. 2004. Seasonal Hunger: Implications for Food and Nutritional Security. In : M.S. Swaminathan & P. Medrano (eds), *Towards Hunger Free India: From Vision to Action*: East West Books Pvt. Ltd., Madras.

Ramachandran, N. 2005. Why Women Must be a Central Concern in Food and Nutritional Policy. *Indian Social Science Review*, 7.

Smith, L., U. Ramakrishnan, A Ndiaye, L. Haddad & R. Martorell. 2003. *The Importance of Women's Status for Child Nutrition in Developing Countries*. IFPRI Research Report 13. Washington, DC: IFPRI. Available at: http://www.ifpri.org/pubs/abstract/131/rr131ch08.pdf.

Trenchard, R. & B. Shreshta. 2002. The Value of Resource Access Learning from Poor Rural Women in Nepal . Paper presented at the WB Regional Workshop on Land Issues in Asia, 4-6 June. Phnom Penh.

UNDP 1998. *Sri Lanka: National Human Development Report 1998*. Regional Dimensions of Human Development, UNDP.

UNCT 2000. *Bhutan: Moving Toward a Common Understanding of the Key Development Challenges in the Country, Common Country Assessment*.

UNDP (various years) *Human Development Report*. Oxford University Press, New York.

World Bank 2004. *Attaining the Millennium Development Goals in India: Role of Public Policy and Service Delivery, Human Development Unit*. World Bank, Washington, DC.

SECTION-7
DRUGDERY AND GENERAL PERSPECTIVES

36

Women in Agriculture Some Issues and Implications

NATARAJU, M.S., R. VINAY KUMAR
AND B. DHANANJAYA

ABSTRACT

Farm women are the backbone of Indian agriculture. They contribute more to food production and family income in terms of skilled and entrepreneurship than is generally known or accepted. Thus the governmental efforts to reach the development goals may be jeopardised by not taking into account the women's contribution to agriculture. In this age of scientific development farm women have not participated to the extent they are expected in agriculture because they are not recognized as tangible contributors to agricultural development. They always remained as invisible workers in spite of their pivotal role and remain silently behind all the efforts of men. Very few scientific and empirical investigations have been made to examine the actual participation of farm women in crop production and dairy production activities. Taking the above situation into consideration, four minuscule studies were conducted in Bangalore, Mysore, Mandya and Chamarajanagar districts of Karnataka to know the level, extent and nature of participation of women in home, farm and livestock production activities. The investigation revealed that majority of farm women regularly participated in almost all the home activities, farm operations and livestock production activities. Rural women's participation in terms of doing was more than that of supervision in case of all agricultural and animal husbandry activities. It is high time for planners, policy makers and extension personnel to realize the immense potential of farm women resources in agriculture and develop appropriate policy measures and mechanical devices to reduce drudgery. Besides organizing effective need based educational programmes to rural women it is necessary to provide critical inputs, credit facilities and market linkages to take up self-employment so as to get additional income and employment opportunity in the days to come.

INTRODUCTION

Women constitute half of the work force in agriculture especially in developing countries. In India according to 2001 census female population was 48.27 percent out of which 72.72 percent came from rural areas. Female labour force participation or more specifically women's participation in agriculture is a much researched and discussed subject. Women are carrying out a number of activities inside and outside the home. This is more true in case of farm women, who work along with men in performing several farm operations as well as performing all the roles related to home, family care and management. Farm women play a significant role in agricultural activities such as seeding, transplanting, weeding, fertilizer application, plant protection, harvesting, processing, storage etc., Several of these operations are exclusively carried out by women only. Their involvement in agriculture varies from region to region and also among different ecological zones, farming systems, caste, class and socio-economic status of families. Thus, by participating in agricultural activities, they directly or indirectly influence the course of agriculture.

Although, a large number of studies have been conducted to identify the roles, needs and problems of women in agriculture, it seems that they have not been sufficient. It is believed that lack of accurate information on women's role in agricultural production has resulted in the failure of development programmes and projects, because they did not address the needs of farm women. An accurate reflection of the contribution of women farmers at village or community level, will further provide the justification for fundamental changes in policies, plans and allocation of agricultural resources so that all farmers, both female and male, would be benefited. This would effectively remove many of the constraints faced by women farmers in increasing their productivity. Apart from this, study on participation of rural women in crop and livestock farming has an important implication for agricultural researchers and extension personnel. With this broad frame of reference, four mini-scale studies were undertaken in Karnataka with the following objectives.

OBJECTIVES

1. To study the level of participation of farm women in farm and home activities
2. To investigate the extent of participation of farm women in farm and home activities
3. To find out the nature of participation of farm women in crop and animal production activities
4. To explore the time utilisation pattern of farm women for household and farm activities

METHODOLOGY

Four mini-scale studies were conducted during 2004-05 and 2005-06. Random sampling technique was adopted to select 210 respondents from four villages of Bangalore, Mandya, Mysore and Chamarajanagar districts of Karnataka representing different agro-climatic zones. The data were gathered using pre-tested structured interview schedule by personal interview method. The collected data were tabulated and analysed using suitable statistical tools. The salient findings of the study are discussed here under.

RESULTS

1. *Participation of rural women in farm and home activities:* Table 1 provides the information on the degree of participation of rural women in paddy cultivation in Mysore district of Karnataka state. It is observed that farm women participated to a very large and considerable extent in 11 out of 17 farming operations. Except the ploughing operation, they have participated in all other activities and to a very large extent in some of the crucial operations such as sowing/transplanting (86%), weeding (84%),

storage of grains (78%), land preparation (72%), cleaning seeds for sowing (70%), gap filling (68%), manure and fertilizer application (68%), harvesting (64%), threshing-winnowing (62%) and rodent control practices (58%). However, their participation was very less in ploughing, seed treatment and marketing.

The table further indicates that more than 50 per cent of the farm women performed domestic chores like cooking and serving, bringing water, preparation of non-perishable foods, grinding flour, cleaning of utensils and home, washing clothes, child care and fuel collection. Further, it was noticed that participation of farm women in activities like shopping, educating children and family budgeting was the least.

Table 1. Participation of rural women in farm and home activities (n = 50)

S. No.	Activity	Number	Per cent
	A. Farm Activities		
1.	Sowing / transplanting	43	86
2.	Weeding	42	84
3.	Storage of grains	39	78
4.	Land preparation	36	72
5.	Cleaning seeds for sowing	35	70
6.	Gap filling	34	68
7.	Manure and Fertilizer application	34	68
8.	Harvesting	32	64
9.	Threshing and Winnowing	31	62
10.	Rodent control	29	58
11.	Preparation of manure and compost	24	48
12.	Bird scaring	21	42
13.	Marketing and transportation	16	32
14.	Labour supervision	14	28
15.	Plant protection measures	13	26
16.	Seed treatment	07	14
17.	Ploughing	0	0
	B. Domestic Activities		
1.	Cooking and serving	46	92
2.	Bringing water	43	86

Contd...

Table 1. Contd...

3.	Preparation of non-perishable food	42	84
4.	Grinding flour	41	82
5.	Cleaning of utensils	40	80
6.	Cleaning home	37	74
7.	Washing clothes	37	74
8.	Child care	35	70
9.	Fuel collection	28	56
10.	Shopping	15	30
11.	Educating children	14	28
12.	Family budgeting	08	16

2. *Extent of participation of farm women in farm and domestic activities*: Table 2 unveils the fact that the largest percentage of farm women participate in sowing (93.3%), followed by storage (88.3%), harvesting (86.6%), scaring birds (81.6%), weeding (80.0%) and threshing (70.0%) followed by fertilizer application (66.6%), land preparation (63.3%), irrigation (63.3%) and plant protection operations (51.6%).

Table 2. Extent of participation of farm women in farm and home activities (n=60)

S. No.	Activity	Total	Extent of participation		
			Regularly	Occasionally	Never
	A. Farm Activities				
1	Sowing	56	42	14	04
2	Storage	53	41	12	07
3	Harvesting	52	37	15	08
4	Scaring birds	49	20	29	11
5	Weeding	48	38	10	12
6	Threshing	42	30	12	18
7	Fertilizer application	40	28	12	20
8	Land preparation	38	28	10	22
9	Irrigation	38	26	12	22
10	Plant Protection	31	20	11	29
	B. Home Activities				
1	Cooking and serving	53	40	13	17
2	Bringing water	50	38	12	17

Contd...

Table 2. Contd...

3	Cleaning of utensils	42	30	12	18
4	Cleaning home	40	35	05	20
5	Washing clothes	40	30	10	20
6	Child care	38	25	13	12
7	Fuel collection	32	20	12	28
8	Shopping	18	08	10	42
9	Children's Education	17	10	07	43
10	Family budgeting	09	06	03	51

Further, it could be observed that large percentage of rural women 'regularly' participated in sowing, storage, weeding and harvesting activities and 'occasionally' involved in scaring birds, harvesting and sowing operation

In addition, farm women were also involved in cooking and serving (88.3%), bringing water (83.3%), cleaning utensils (70.0%), home (66.6%) and clothes (66.6%), child care (63.3%) and fuel collection (53.3%); while less number of them were involved in family budgeting (15.0%), children's education (28.3%) and shopping (30.0%).

3. *Nature of participation of farm women in agricultural activities*: It is interesting to note from Table 3 that of the twenty agricultural activities women were less involved compared to men in seven operations namely, preparatory tillage, nursery bed preparation, forming ridges and furrows, digging pits, plant protection, irrigation and account keeping. Because of greater physical efforts needed for these activities, men are more suited for these operations in the field. Women were more involved in other farm operations like weeding (96%), scaring of birds (96%), harvesting (94%) and post-harvest activities like threshing (92%), grading (98%), winnowing (90%) and storage activities (94%), followed by sowing (92%), transplanting (94%), seed treatment (90%) and fertilizer application (92%).

   After establishing the pattern of participation by farm women, the next aim of the research was to analyse the nature of participation. The nature of participation of women refers to doing and supervising. In general, women's participation in terms of doing was more than supervision. Among the farm women, the percentage of those doing ranged from a minimum of about eight per cent in preparatory tillage to a maximum of about 96 per cent in grading seeds and other farm produce. Regarding supervision of practices, the percentage of women ranged from the minimum of about two per cent in grading activity to the maximum of about 26 per cent in plant protection preceded by seed treatment (22%), digging of pits (20%), preparatory tillage (20%) and nursery bed preparation (20%).

Table 3. Nature of participation of farm women in agricultural activities (n = 50)

S. No.	Activity	Supervision		Doing		No participation	
		No.	Per cent	No.	Per cent	No.	Per cent
1	Preparatory tillage	10	20	04	08	36	72
2.	Nursery bed preparation	10	20	12	24	28	56
3.	Forming ridges and furrows	04	08	11	22	35	70
4.	Digging pits	08	16	13	26	29	58
5.	Seed treatment	11	22	39	78	10	20
6.	Sowing	04	08	42	84	04	08
7.	Fertilizer application	08	16	38	76	04	08
8.	Transplanting	07	14	40	80	03	06
9.	Weeding	02	04	46	92	02	04
10.	Scaring of birds	02	04	46	92	02	04
11.	Watch and ward	07	14	39	78	04	08
12.	Plant Protection	09	18	16	32	25	50
13.	Irrigation	13	26	12	24	25	50
14.	Harvesting	02	04	45	90	02	04
15.	Threshing	08	16	38	76	04	08
16.	Winnowing	04	08	41	82	05	10
17.	Grading	01	02	48	96	01	02
18.	Marketing	08	16	32	64	10	20
19.	Storage	02	04	45	90	03	06
20.	Account keeping	02	04	22	44	26	52

4. *Nature of participation of farm women in livestock activities*: Women's participation in the area of livestock keeping was studied with reference to ten specific areas as given in Table 4. Unlike their limited direct participation in agricultural activities, majority of women were involved in almost all the livestock activities namely, feeding animals (96%), fodder collection (96%), maintenance of cattle shed (94%), cow dung making for kitchen fuel (94%), milking (88%), marketing of milk and milk products (84%), cleaning of cattle (80%), health care of animals (80%), purchase of cattle feed and maintenance of accounts (64%). As high as 36 per cent of respondents showed non-participation in maintenance of accounts followed by purchase of cattle feed (34%), health care (20%) and cleaning of cattle (20%). When the nature of participation was analysed, it was found that in general, women's participation in terms of doing is more than in terms of supervision.

A high percentage of farm women under 'doing' were found in activities like cow dung making (88%), fodder collection (84%), maintenance of cattle shed (80%), feeding animals (78%), followed by other practices; whereas, large percentage of farm women supervised the following activities: health care (36%), cleaning of cattle (24%), purchase of cattle feed (24%), maintenance of accounts (20%), milking (18%), feeding of animals (18%), maintenance of cattle shed (14%), fodder collection (12%), marketing of milk (8%) and cow dung making (6%).

Table 4. Nature of participation of farm women in livestock activities (n = 50)

S. No.	Activity	Supervision		Doing		No participation	
		No.	Per cent	No.	Per cent	No.	Per cent
1.	Milking	09	18	35	70	06	12
2.	Marketing of milk and other dairy products	04	08	38	76	08	16
3.	Feeding animals	09	18	39	78	02	04
4.	Maintenance of cattle shed	07	14	40	80	03	06

Contd...

Table 4. Contd...

5.	Fodder collection	06	12	42	84	02	04
6.	Cleaning of cattle	12	24	28	56	10	20
7.	Cow dung making	03	06	44	88	03	06
8.	Health care	18	36	32	64	10	20
9.	Maintenance of accounts	10	20	22	44	18	36
10.	Purchase of cattle feed and other items	12	24	21	42	17	34

5. *Time utilisation pattern of rural women* in Bangalore rural district is given in Table 5. An an average, rural women spend 38 per cent of their time on household activities, 22 per cent on agricultural activities and 40 per cent on sleeping, resting, social and leisure time activities. Among different economic categories, farm women of middle and high income category spent more time on household activities (40.7%) compared to rural women of low income category (23.6%). Similar trend was observed with respect to per cent utilisation of time on sleeping, resting and leisure time activities by three categories of respondents, viz., high economic group (56.7%), middle economic group (41.5%) and low economic group (33.3%). It was also found that the low economic category rural women spent more time per day (34.1%) in agricultural activities compared to farm women of middle (18.8%) and high income categories (12.5%).

   All the low economic category rural women respondents were engaged in paid agricultural work and none were doing their own farm work. Low economic category rural women spent more time in dish washing as they did not have servants to assist them. Further, it was observed that medium and. high categories of respondents spent more time on household activities and allied agricultural activities like poultry, dairy, kitchen gardening etc., while low economic category of women spent more time on agricultural activities in order to earn their livelihood.

Table 5. Time utilisation pattern of rural women

S. No	Activity	Average time spent (hrs./day)							
		Low income category		Middle income category		High income category		Mean	
		Hrs.	Per cent	Hrs.	Per cent	Hrs.	Per cent	Hrs.	Per cent
1	Food preparation	2.8	11.5	3.2	13.3	2.1	8.7	2.7	11.3
2.	Dish washing	0.6	2.5	0.7	2.8	0.4	1.7	0.6	2.5
3.	Cleaning of home	0.7	2.8	0.9	3.8	0.8	3.4	0.8	3.3
4.	Cleaning of clothes	0.5	2.1	0.5	0.8	0.2	0.8	0.3	1.3
5.	Marketing	0.7	2.8	0.5	2.1	0.2	0.8	0.5	2.1
6.	Family care	1.5	6.3	2.1	8.7	2.9	12.1	2.2	9.2
7.	Personal care	1.0	4.6	2.2	9.2	3.2	13.3	2.0	8.3
8.	Sleeping and resting	6.8	28.3	7.7	32.2	8.1	33.8	7.5	31.3
9.	Leisure time	1.2	5.0	2.0	8.3	3.1	12.9	2.1	8.7
10.	Agriculture	7.3	30.4	3.1	12.9	0.8	3.3	3.8	15.8
11.	Allied agricultural	0.9	3.7	1.4	5.9	2.2	9.2	1.5	6.2
	Total	24.0	100.0	24.0	100.0	24.0	100.0	24.0	100.0

CONCLUSION

The studies have brought out several realities about farm women. Majority of the farm women 'regularly' participated in almost all the home activities, farm operations and livestock production activities. Rural women's participation in terms of 'doing' was more than that of supervision in case of agricultural and livestock production activities. Farm women of middle and high income category spent more time on household activities compared to low income group.

IMPLICATIONS

1. It is necessary to conduct location-specific research which generate more concrete and detailed data that would be more effective and practical from developmental point of view.
2. Efforts should be made to develop appropriate techniques whose adoption will reduce the drudgery of farm women and improve their efficiency and physical input.

3. Education programmes should be organised after assessing the training needs of different categories of farm women. Adequate attention should be given to adult education programmes by way of organising training in crop production, livestock management and other income generating activities viz., tailoring, and crafts at convenient time and place based on their needs.

4. Provision of appropriate crop and livestock technology, inputs and services, training and education, organised marketing and credit and participation in cooperative institutions will result in raising the productivity, employment and income potential of rural women and in improving their socio-economic status and quality of life.

5. Women have not been provided easy access to extension training in appropriate technology, services, inputs and credit facilities. Institutions like cooperatives, credit agencies should make efforts to encourage women through adult education programmes.

37

Ergonomic Evaluation of Cutting and Uprooting Activity Performed by Farm Women from Hilly Regions

JATINDER KISHTWARIA, ARUNA RANA
SHIPRA SOOD & SUMAN LATA

ABSTRACT

The present paper examines the effect of fodder cutting and uprooting activity on the cardiovascular system of farm women. The research design comprised of field study conducted on 30 farm women. The activity of cutting and uprooting were studied with the traditional tools as well as with the improved tools. Newly designed and developed slashing knife and uprooter were provided to the respondents for performing the activity of cutting and uprooting. The results revealed significant reduction in average working heart rate (2.08** - cutting and 4.99** - uprooting; Significant at 5% level) and peak heart rate (-3.55**- cutting and 1.70**- uprooting; Significant at 5% level). It was found that there was significant reduction in muscular stress after use of improved tools over the existing tools.

INTRODUCTION

It is believed by some historians that it was women who first initiated farming by domesticating wild crop plants. While men went out hunting, women started gathering the seeds of plants and began cultivating them to meet their food, fibre and fuel needs (Swaminathan, 1985). In the developing countries, women in the rural areas are often considered as 'invisible farmers'. Besides engaging themselves in a variety of activities, both on the farm and at home, the farm women also contribute to family income through their wage earnings. Their earnings form a major part of the income of poor households. Poorer the farm household, greater is women's relative contribution to its total income. They contribute a larger share of what they earn to basic family maintenance than men (Mencher, 1987). Majority of the economically active women are from the marginal and small holder categories (Duvvury, 1989). Many agricultural operations and household activities performed by women involve a lot of physical strain which create serious health problems in the long run. Since they are overburdened with so much work on the farm and at home, they continue to neglect their health (Vatsala, 1993). Information on various kinds of health hazards, in activities where farm women are engaged or are exposed, are very limited. Therefore the present study was undertaken to carry out the ergonomic assessment of women performing cutting and uprooting activity.

METHODOLOGY

The study was carried out in Kangra District of Himachal Pradesh. The ergonomic assessment was done on a total sample of 30 farm women performing the activity of cutting and uprooting.

Assessment of ergonomic cost: Assessment of the health status of hill farm women along with the ergonomic cost of cutting and uprooting activity was studied. The parameters of health status included body weight, height, body temperature, physical fitness index, lean body mass, body composition and oxygen consumption.

The variables of ergonomic cost of work studied were: heart rate (recorded with the polar heart rate monitor), energy expenditure [0.159 heart rate (bpm) - 8.72], total cardiac cost of work [CCW+CCR where Cardiac cost of work (CCW) is Average working heart rate-Average resting heart rate], Duration of activity, and Cardiac cost of recovery (CCR) (Average recovery heart rate-Average resting heart rate), Duration of activity, physiological cost of work (TCCW/ Total time of work) and postural analysis.

RESULTS

Table 1: Physiological work load while performing cutting and uprooting using existing and improved tools

Physiological parameters	Traditional Tools		Improved Tools		Significant change (t test)	
	Cut	Upr.	Cut	Upr.	Cut	Upr.
Average working heart rate (Beats/min.)	110	125	106.53	116	3.15 ↓ (2.08**)	7.2↓ (4.99**)
Average peak heart rate (Beats/min.)	121	132	117.7	127.76	2.72 ↓ (-3.55**)	3.21↓ (1.70**)
Average energy expenditure (Kj/min.)	8.78	11.28	8.21	9.71	6.49↓	13.91↓
Peak energy expenditure (Kj/min.)	10.55	12.30	9.99	11.58	5.30↓	5.85↓
Average TCCW (Beats)	809	531	621	317	23.23 ↓	40.30↓ (4.47**)
Average PCW (Beats/min.)	26.95	53.1	21.72	31.66	19.40↓	40.37↓
Average RPE	4	5	2	3	50↓	40↓
Output Parameters						
Total time taken for the activity (min.)	30	10	30	10	-	-
Coverage of land under operation (Sq.mt.)	22.94	11.95	57.77	16.04	-151.83↓ (2.67**)	34.22 ↓ (0.48)
Production per unit (kg)	15.75	9.88	25.23	11.3	-60.19↓ (0.91)	-14.67 ↓ (1.24)
Environmental parameters						
Mean temp. °C	26.76	27.13	29.06	30.92		
Mean relative Humidity%	76.24	76.4	68.53	65.33		

* Significant at 1 tail ** Significant at 2 tail

The ergonomic cost of cutting and uprooting was calculated for thirty rural farm women of Kangra District of Himachal Pradesh. The ergonomic cost was studied in terms of average and peak heart rate, average and peak energy expenditure, total cardiac cost of work(TCCW) and physiological cost of work (PCW). According to

AICRP report (1996-2001) heart rate and energy expenditure were highest for uprooting of seedling, wheat harvesting and bundling.

It is clear from Fig. 1&2 as well as table 1 that the average and peak heart rate values and average peak energy expenditure values were higher than the acceptable limits with the use of both the traditional and the improved tools. Significant reduction was observed in both cutting (Significant reduction in AHR-3.15↓) and uprooting (Significant reduction in AHR-7.2↓) activity when carried out with the improved tools. The heart rate values were more than the permissible limits because cutting and uprooting is a very tedious work and have to be carried out in the traditional methods. By use of new improved tools the total cardiac cost of work and physiological cost of work being a derived parameter of heart rate values showed significant reduction in both TCCW (Cutting-23.23 ↓; Uprooting-40.30↓) and PCW (Cutting-19.40 ↓ and Uprooting-40.37↓). T test was also applied to test the significant difference between the two tools. Significant difference was seen in both cutting (2.08**) and uprooting (4.99**) activity when the performance was compared with the traditional and the improved tools.

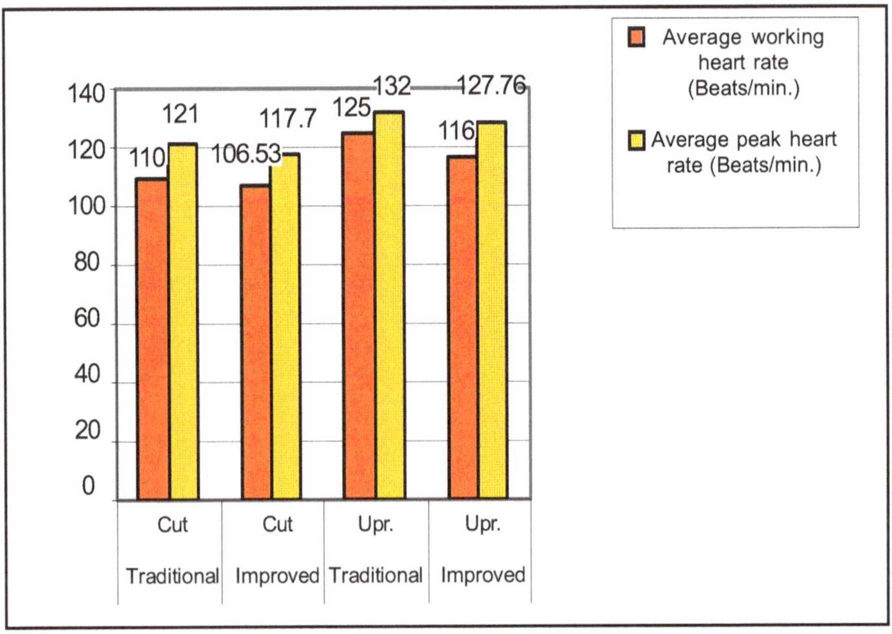

Fig. 1. Average and Peak Heart Rate of Women Respondents While Working

In the output parameters, significant increase was observed in coverage of land under operation (151.83 ↓ in cutting and 34.22 ↓ in uprooting) thereby increasing the production too while working with the improved cutting and uprooting tools. The mean temperature and relative humidity while cutting of maize with the improved tool was 29.06^{0}C and 68.53% and the same during uprooting activity with the improved tool was 30.92^{0}C and 65.33%.

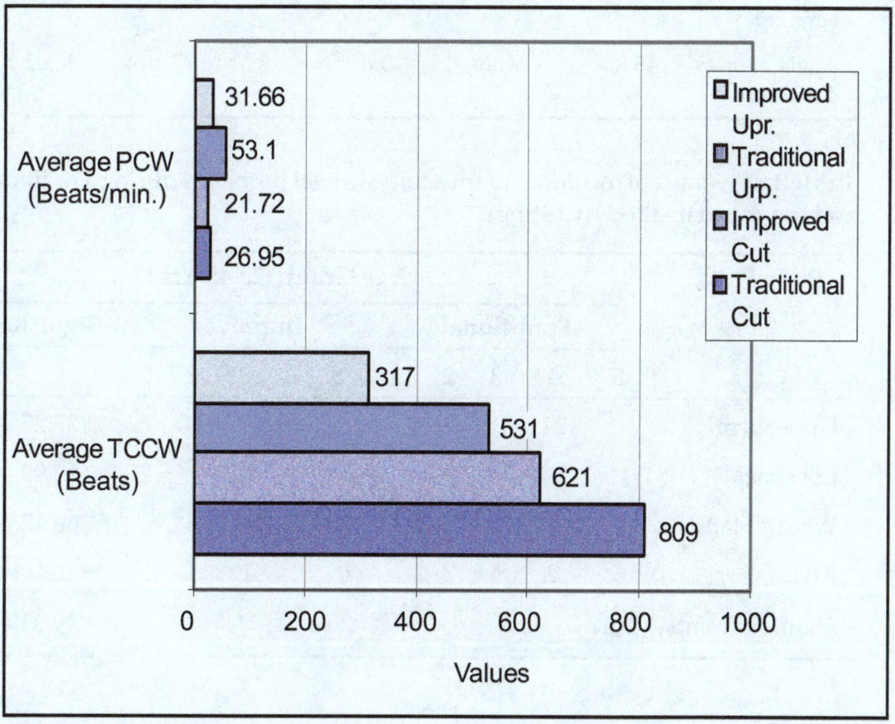

Fig. 2. Total Cardaic Cost and Physiological Cost of Farm Women While Working

The body angles were measured with the help of a flexi curve. Only standing and bending postures were used for performing the activity of cutting and uprooting. As seen in table 2, there has been much reduction in angle of deviation (3.6^0 in cutting and 5.5^0 in uprooting with improved tool as compared to 15.86^0 in cutting and 7.86^0 in uprooting with traditional tool) thereby reducing the body pain and also minimizing the muscular stress.

Table 2. Postural analysis (Angle of bend)

Postural analysis	Traditional Method		Improved		Percent change in AOB	
	Cutting	Uprooting	Cutting	Uprooting	Cutting	Uprooting
Angle of normal curve	192.14^0	192.14^0	191.6^0	191.6^0	-	-
Angle while bending	208^0	200^0	195.2^0	197.1^0	-	-
Angle of deviation	15.86^0	7.86^0	3.6^0	5.5^0	77.30 ↓	30.02 ↓

Table 3a. Average of incidence of musculo-skeletal problems during Traditional and Improved methods (Cutting)

Body Parts	Age Group (20-45 Yrs.)										
	Traditional					Improved					Signi. Red.
	5	4	3	2	1	5	4	3	2	1	
Upper arm		21		2					10		77.27 ↓
Low back	15	3						10		2	61.90 ↓
Wrist/ Hands		10		11				3	6		66.12 ↓
Knees		2	6	8							100 ↓
Shoulder Joints	30							4		2	89.33 ↓

The table 3a depicts that there has been significant reduction in pain in upper arm (77.27% ↓), low back (61.90 % ↓), wrist/hands (66.12 % ↓), shoulder joints (89.33 % ↓). Cent percent reduction in pain was reported in knees while performing the cutting activity. Some of the occupational health hazards of farm women engaged in agriculture (Suman & Maurya, 1993) include: Cuts, wounds and injuries; allergy, skin and eye irritation; swollen and sore hands and feet; chest congestion and breathing problem; body ache; sunstroke; physical tiredness, fatigue and exhaustion; psychological fatigue; food poisoning; malnutrition and anaemia.

Table 3b. Average incidence of musculo-skeletal problems during Traditional and Improved methods (Uprooting)

Body Parts	Age Group (20-45 Yrs.)										
	Traditional					Improved					Signi. Red.
	5	4	3	2	1	5	4	3	2	1	
Upper arm		9	13				3		8		60.56 ↓
Wrist/ Hands	26							2	6		86.15 ↓
Knees		10		5			2			5	58 ↓
Ankles/ Feet		20	2					7			75.58 ↓
Shoulder Joints	25	3					3			11	73.22 ↓

The data in table 3b clearly indicates the significant reduction in pain in wrist/hands (86.15 % ↓) followed by ankles/feet (75.58 % ↓), shoulder joints (73.22 % ↓), upper arm (60.56 % ↓) and knees (58% ↓).

CONCLUSIONS

- The use of improved tools for carrying out the activity of cutting and uprooting showed significant reduction in average and peak heart rate.
- Significant reduction was observed in both the total cardiac cost of work and physiological cost.
- Output parameters showed significant increase in land covered under operation and production per unit.
- Significant reduction in musculo skeletal problems was found while working with the improved tools.

REFERENCES

AICRP report (1996-2001) All India Coordinated Research Project on Home Science, Indian Council of Agricultural Research, New Delhi. 114pp.

Duvvury, N. 1989. Women in agriculture: a review of the Indian literature. *Economic and Political Weekly* 24(43): WS 96-112.

Mencher, J. 1987. Women's work and poverty: women's contribution to household maintenance in two regions of South India. In: *A Home Divided: Women* and

Income Control in the Third World. (Eds.) Deoyer, D. & J. Bruce, Stanford University Press, Stanford, USA.

Suman, P. M. & R. P. Maurya. 1993. Health status and occupational hazards of farm women. Paper presented at the *National Seminar on Women in Agriculture - Developmental Issues*, December 28-30, National Academy of Agricultural Research Management, Hyderabad, India.

Swaminathan, M. S. 1985. In: *Report from the CGIAR Inter-Centre Seminar on Women and Agricultural Technology*, Bellagio, Italy, March 25-29, Rockefeller Foundation and ISNAR, Vol. I, p. 51- 73.

Vatsala, R. 1993. Health hazards of women in agriculture. Paper presented at the *National Seminar on Women in Agriculture - Developmental Issues*, December 28-30, National Academy of Agricultural Research Management, Hyderabad, India.

38

Ergonomic Assessment of Farm Women Performing Weeding and Winnowing Operations

DEEPALI CHAUHAN & REKHA DAYAL

ABSTRACT

The present study aims to assess postural and muscular stress among farm women performing weeding and winnowing operations. Postural stress was assessed through postural deviation in spinal curvature, energy expenditure, physical cost of work, cardiac cost of work, pain in different body parts and muscular stress was assessed by percentage decrease in grip strength. Postural deviations were recorded with the help of flexi curve. Heart rate was calculated in 3 phases: before, during and after the work. Energy expenditure, physical cost of work, cardiac cost of work was also calculated. Grip strength of both hands was calculated by grip dynamometer. The mean value of ergonomic parameters were found more in weeding operation which were as follows : -Postural deviation (25^0), Average heart rate (5.35), energy expenditure (5.70), physical cost of work (3.05), total cardiac cost of work (35.82). After weeding operation farm women perceived very severe pain in lower back, buttocks, and palms, while after winnowing they experienced very severe pain in shoulder, upper back and upper arms. Percentage decrease in grip strength was found more in weeding operation than winnowing operation. The reason behind it may be that weeding is performed in squatting posture with sustained walking whereas winnowing was performed in standing posture keeping both hands in upward direction without any support. Both the activities come under the category of moderately heavy tasks. There is great need for farm mechanization by devising appropriate tools and techniques that can eliminate awkward body positions and awkward body movements. Finally it will decrease work stress and increase the work efficiency, health and body comfort of farm women.

INTRODUCTION

Agriculture is the largest industry in India and women participate in almost all agricultural operations such as transplanting, weeding, threshing etc. Women constitute fifty percent of the agricultural work force in India giving India the highest degree of female participation in agriculture (Tamiliselivi & Undamn, 2000). The nature of agricultural work is diverse. It is physically demanding, involves external time constraints and experiences rapid changes in environment and tasks. Rigorous nature of farm work exposes worker to a number of risk factors that have been associated with musculo-skeletal disorders. Heavy lifting, working in awkward positions for a prolonged period of time and poorly designed tools and implements take a toll on farm women and result in musculo-skeletal problems, the most commonly reported health problem. In addition to farm work, rural women have to perform household tasks as well as attend to livestock. Triple burden of household, livestock and farm work make the farm women more vulnerable to musculo-skelatal disorders. The application of ergonomic principles facilitates adjustments to the working conditions and job demands according to the capabilities of working populations. The goal of ergonomics is to design facilities, furniture, equipments, tools, work process and job demands to be compatible with human capacities and limitations. In the present study, therefore, an attempt was made to find out the effect of two selected activities, weeding and winnowing on anatomical and physiological stress in female farmers.

MATERIAL & METHODS

The study was conducted at student farm of C.S.A.U.A & T, Kanpur. Fifteen physically fit farm women were selected for the study. Postural deviations in lumber region of farm women was assessed through flexi curve, time spent in different postures was recorded with the help of stopwatch, musculo-skeletal problems were measured through 5 point scale and stethoscope was used for measuring heart rate. Heart rate was measured in two phases: split up cycle and complete cycle. At each stage heart rate was measured 4 times for the purpose of accuracy. Cardiac cost of work, energy expenditure and physical cost of work were measured. Muscular stress was measured through grip dynamometer by assessing percentage change in grip strength of both the hands.

Table I. Postural deviation in lumber region of farm women.

Postural Deviations			
Posture		Angle of Deviation	
Weeding	Winnowing	Weeding	Winnowing
Squatting	Standing	25^0	20^0

Table 2. Mean value of total time spent by farm women in different postures during weeding and winnowing

S.No	Body parts	Posture	Total time spent (hours)	
			Weeding	Winnowing
1.	Neck	1. Free and relaxed	0.314	0.871
	Shoulder	2. Tense due to work	3.686	3.219
2.	Elbow	1. Free in posture of choice	0.340	0.187
		2. Arms in positions as required by the work.	1.57	1.573
		3. Arms tense due to work or joints in an extreme position	2.09	2.07
3.	Back	1. In a natural posture.	0.152	0.289
		2. In a good posture but limited by work.	1.678	2.026
		3. In a stooped position Bend/or/ and poorly supported	1.810	1.685
4.	Legs	1. In a free position	0.103	1.013
		2. Legs forward	2.058	1.563
		3. Legs backward	1.839	1.424

RESULTS AND DISCUSSION

1. *Assessment of postural stress among farm women:* Table1 shows that during weeding operations postural deviation in lumber region of the farm women was more than in winnowing operation.

   From table 2 it is explicit that during weeding operation, neck and shoulders, elbows and back of farm women were found tense due to work for longer period than during winnowing operation. During both activities, legs were kept in forward position for most of time but in weeding operation, time spent by legs of farm women in forward position was more than in winnowing operations.

Table 3. Distribution of the farm women as per feeling of pain in body parts during weeding and winnowing operations (W_1 = weeding: W_2 = winnowing).

Body Parts	Very Severe		Severe		Moderate		Low		Very Low	
	W_1	W_2	W_1	W_2	W_1	W_2	W_1	W_2	W_1	W_2
Neck	–	1 (2%)	25 (50%)	7 (14%)	14 (28%)	25 (50%)	9 (18%)	12 (24%)	2 (4%)	5 (10%)
Shoulder	–	37 (74 %)	6 (12%)	12 (24%)	38 (76%)	1 (2%)	5 (10%)	–	1 (2%)	–
Upper back	–	48 (96%)	–	28 (56%)	45 (90%)	3 (6%)	5 (10%)	–	–	–
Upper Arms	5 (10%)	–	30 (60%)	2 (4%)	15 (30%)	–	38 (76%)	–	–	–
Mid Back	–	40 (80%)	–	–	6 (12%)	2 (4%)	2 (4%)	3 (6%)	6 (12%)	45 (90%)
Lower Arms	21 (42%)	28 (56%)	27 (54%)	10 (20%)	–	–	–	–	–	–
Lower Back	30 (60%)	–	18 (36%)	22 (44%)	2 (4%)	–	–	–	–	–
Buttock	46 (92%)	32 (64%)	4 (8%)	–	–	3 (6%)	–	26 (52%)	–	21 (42%)
Palms	39 (78%)	18 (36%)	11 (22%)	–	–	–	–	–	–	–
Fingers	18 (36%)	28 (56%)	12 (24%)	15 (30%)	–	15 (30%)	3 (6%)	5 (10%)	–	2 (4%)
Thigh	10 (20%)	27 (54%)	25 (50%0)	4 (8%)	10 (20%)	4 (8%)	3 (6%)	10 (20%)	–	2 (4%0)
Legs	3 (6%)	15 (30%)	27 (54%)	21 (42%)	17 (34%)	21 (42%)	3 (6%)	10 (20%)	--	2 (4%)

2. *Assessment of musculo-skeletal problems among farm women:* According to the data shown in table 3, during weeding, very severe pain was felt in buttocks by maximum number of farm women (92%) and severe pain was felt in upper arms by 60 percent.

   On the other hand, it is also shown by table 3, during winnowing operation very severe pain was felt in upper back by a maximum number of farm women (96%) and severe pain was felt in upper back by 56 percent farm women.

3. *Assessment of Average heart rate (A.H.R), energy expenditure (E.E), cardiac cost of work (C.C.W) and total cardiac cost work(T.C.C.W) during 3 phases of the work :* From table 4, it is clear that mean value of different parameters of physical strain namely, A.H.R., T.C.C.W, P.C.W. and E.E. were found highest during the work than during onward and backward journey of the split–up cycle of weeding and winnowing operation.

   According to table 4, the mean value of these parameters were greater for weeding operation than winnowing operation. It is also evident that during onward and backward journey the mean value of these parameters were almost similar for both the activities namely, weeding and winnowing operations.

Table 4. Mean value of different parameters of physical strain during 3 phases of weeding and winnowing operation.

Particulars	A.H.R.		T.C.C.W	P.C.W	E.E	
	Weeding	Winn-owing	Weeding	Winn-owing	Weeding	Winn-owing
On ward journey	5.35	5.38	35.55	35.47	6.75	5.9
At the work.	38.18	33.43	243.4	222.8	9.53	8.83
Backward journey	6.75	6.03	38.93	38.07	5.26	0.80

*A.H.R = Average Heart rate, T.C.C.W. = Total cardiac cost of work, P.C.W. = Physical cost of work, E.E. = Energy expenditure

On the basis of the results of table 1, 2, 3 and 4, it can be concluded that the level of postural stress was more during weeding than winnowing operation. It was because of greater angle of deviation in lumber region of spine from the normal body curve, most of time spent in tensed working posture thus resulting in higher values of different parameters of physical strain during weeding.

4. *Assessment of muscular stress among farm women:* Table 5 reveals that percentage change in grip strength of both hands was found more during winnowing operation. Positive sign showed that grip strength was increased after the activity. It may be due to the warming up of muscles at the start of activity. Bimla *et al.*, (2001) also observed that grip strength increased up to 9-6 percent in the first cycle of picking cotton whereas it decreased by 14.2 percent at the end of the activity being done for above 5 hours. It is obvious from the results that grip muscles were used more during winnowing operation than in weeding operations. Due to this reason musculo-skeletal problems were found more in upper arm and upper back of farm women after the winnowing operations.

Table 5. Percentage change in grip strength (kgs.) of the subjects after weeding and winnowing activity.

Activities	Percentage change in grip strength	
	Left	Right
Weeding	+ 9.98	+ 15.99
Winnowing	+ 17.45	+18.23

- No Percentage change in grip strength was found during onward and backward journey of weeding and winnowing operation.

On the basis of findings of the study, it was evident that both weeding and winnowing were strenuous activities but weeding was some what more strenuous than winnowing due to the squatting posture adopted during the activity. Biomechanical research shows that high spinal compression occurs in squatting/stooped postures and that sustained or repeated flexion of the spine may disturb the neuro-muscular stability of the lower back and increase the risk of fatigue leaving the back more vulnerable to injury. Further, it was noted that intradiscal pressure is higher in sitting than standing posture (Nicholson & Edstrom, 1970). Hence there is a great need of ergonomic intervention for reducing the work stress of weeding and winnowing operation in terms of postural and muscular stress of farm women through developing appropriate tool and techniques for creating a fit between physical capacity of farm women and work requirement as well as eliminating the risk factors of weeding and winnowing.

References

Bimala, R., S. Gandhi & M. Dilbaghi. 2001. Ergonomic evaluation of farm women picking cotton. Paper presented at *International congress on humanizing work and work environment*, 2001. Indian Institue of Technology, Mumbai.

Nicholson, A. & C. Edstrom. 1970. Intervital dynamic pressure management in lumber disc. *J. Rehabiltation Med.* 8 (3)

Tamilselvi, D. & S. Undamn. 2000. The extent of adoption of modern technologies among the farm women. *Agri. Ext. Rev.* May – June, Pp. 10-15.

39

Women's Employment Potential Status in *Charaka* Reeling Units in Karnataka: Pattern and Activity Analysis

CHINNASWAMY, K.P., N. KUMARSWAMY
S. FATIMA, K. RASHMI AND N. JAGDISH

ABSTRACT

Sericulture, an agro-based industry contributes significantly to Indian economy. Silk cocoons are reeled on *charaka*, cottage and multiend reeling machines. Of these more than 50% of raw silk comes from *charaka* sector. *Charaka* silk reeling units decreased from 23971 to 16263 (1998-2003) in Karnataka. These *charakas* are mostly operated at grass root level with labour coming from small, marginal and land less villagers. A study was conducted in 15 *charaka* reeling units in two *taluks* each of Bangalore rural, Mysore, Kolar and Chamarajanagar districts. Respondents were grouped into small (1-4 *charakas*), medium (5-8) and large *charaka* reelers (>8). Primary data was collected with pre-tested schedules.

Analysis revealed that, employment generation differed with the size of the reeling units. On an average employment potential were 1022.47,

2711.43 and 3882.38 man-days in small, medium and large *charaka* units/ year respectively. Small *charaka* units employed more family labour (78.01%), compared to large units (28.32%). About 25.48% of women labour comes as hired, while, 16.67% labour contribution is from family. However, more than 40% of women are employed in *charaka* reeling units irrespective of size. Majority (87.4%) of the women are engaged in collection of dead pupae, jute extraction and feeding fuels, followed by cooking, changing of water from pans and reeling operation (50.9%) and 28.40% in wheel operation. None of the women participated in cocoon purchase and silk marketing. Inspite of women forming 42.20% workforce in *charaka* reeling industry, relatively less wage is paid to women. The job is associated with health problems. Providing regular health check-up and other benefits is important. The paper discusses other issues related to the constraints faced by the women involved in silk reeling activities.

INTRODUCTION

Sericulture is an agro-based industry and occupies an important position in Indian economy owing to its development, generation, production and export of silk. Its manufacture comprising of mulberry cultivation, cocoon production, silk reeling and weaving of silk fabrics needs strong infra structure. It has relatively high employment potential and therefore, it would be an ideal sector for rural development in India. Indian silk industry provides gainful employment to over sixty million people directly or indirectly. Major strength of Indian silk industry is it's ability to produce varieties of silk fabric in design, colour and their combinations. Silk has enjoyed a sturdy expanding and profitable market in the country and abroad.

Karnataka is the major traditional silk producing state in India contributing maximum raw silk in the country. The silk industry of the state comprises of 23,776 *charaka* reeling units, 1,522 cottage basins, 21,021 filature basins, 960 multiend units and about 265 automatic reeling units to produce raw silk. Among these, major contribution is from *charaka* reeling units (Anon, 2002).

In India, traditional reeling technique known as "charaka silk reeling", is popular at grass root level among poorer section of society. Production of *charaka* silk accounts for about 50 per cent of the total silk produced in the country. It is fascinating to see such fine quality raw silk produced on such crude *charakas*. Such wonderful yarn is produced by the personnel skill / craftsmanship of silk reelers. Similarly, quality silk fabrics with rich designs are produced on graded handlooms in India. Improved reeling techniques such as cottage basins and multiend reeling were installed to get quality mulberry silk yarn (Sonwalkar *et al.*, 1987).

Charaka system of reeling is very primitive. Presently, around 50 per cent of the silk is produced on *charaka*, while 40 per cent comes from cottage basins and the rest from multiend reeling basins. The main reasons why *charaka* is still dominating Indian silk reeling industry is that the bulk of the raw silk produced is used by handloom sector, wherein the cost factor of raw material has to be kept as low as possible, for the rate of finished product to be saleable. Secondly, the establishment of *charaka* unit does not require big investment and special skill and this enables *charaka* silk to be sold at an affordable price, which is the need of handloom industry. Thirdly, the inferior quality multivoltine and defective cocoons available at lower prices can be reeled more economically on *charaka* than on cottage basin or multiend basins (Anon, 1990).

As per the available data, number of *charaka* units (31.9 %) and quantity of *charaka* silk produced (21.87 %) has decreased over the years (1992-93 to 2002-2003). Some reelers are continuing the *charaka* silk reeling enterprises, while others discontinued either permanently or temporarily, and entered other business. In order to understand the *charaka* silk reeling status as a whole and factors contributing to the sustainability of these sectors, the present study was under taken.

MATERIAL & METHODS

The study was conducted in Karnataka State. A multistage sampling procedure was adopted for the purpose of selection of representative districts and taluks for the survey. Silk reeling is carried thoughtout the state but large-scale reeling is concentrated mainly in southern Karnataka (traditional sericulture area). Hence, to represent fair sampling, four districts in traditional areas viz. Bangalore (rural), Kolar, Mysore, Chamarajanagar having more concentration of *charaka* units were selected. Silk reeling is practiced in most of the *taluks* of selected districts. However, it was focused on *taluks* having large-scale silk operation and having highest reelers in the *charaka* system of reeling. Accordingly Chamarajanagar, Kollegal, Yalandur *taluks* from

Chamarajanagar district, Ramanagar and Chanapattanna *taluks* in Bangalore (rural) district, Chintamani and Sidlghatta *taluks* in Kolar districts and T.Narasipura *taluk* in Mysore district were selected for the study. Total respondents numbered 120. Fiften *charaka* reelers from each *taluk* were selected randomly. Reelers were interviewed with pre-tested schedule for elucidating the data required for the study and analyzed.

RESULTS

Silk reeling is highly labour intensive activity. It's potential in solving the unemployment problem in rural India has been recognized by the Central and State Governments. In this context an attempt has been made to estimate the quantum of employment generated in *charaka* silk reeling units of different unit holders (sizes).

Small size charaka silk reeling units: Purchase of cocoons and sale of raw silk were attended by only male members of the family (171.18 man-days in a year). Most of the male members of the family and women were involved in cocoon cooking and silk reeling (423.07 man-days), whereas the hired labourers were involved in changing of water in the pans and some in cooking and reeling (160.86). Majority of the family members-both male and female (167.17 man-days) were engaged in wheel turning operation compared to hired labour (36.77 man-days). Dead pupae collection, jute extraction and feeding of fuel were attended by family labour (36.26 man-days) and hired labour (34.82 man-days). In small *charaka* reeling units totally 1,022.47 man-days were required per annum; of this 78.01 per cent was constituted by family labour and 21.99 per cent of hired labour was employed in different activities in the *charaka* silk reeling (Table 1).

Medium charaka silk reeling units: The average medium *charaka* silk reeling unit employed 2,711.43 man-days per year (Table 1). In this, purchase of silk cocoons and selling the raw silk were attended by family members (444.75 man-days per year). Silk cocoon cooking, reeling and changing of water in the pans were attended by 1099.2 man-days of family labour and 457.2 man-days by hired labour, Both men and women of family and female hired labour attended to cooking and reeling. Others attended to changing of water in the pans. The next activity i.e. wheel operation was attended by about 434.36 man-days by family labour and 104.53 by hired labour. Dead pupae collection, jute extraction and feeding of fuel were attended by 94.21 man-days by family members and 77.18 man-days was contributed by hired labour. Even in this majority of females in the family and amongst hired labours attended to these activities.

Large charaka silk reeling units: In case of large *charaka* reeling units, 3,882.38-man day were employed per annum. In this 71.61 per cent was hired labour and 28.32 per cent was family labour (Table 1). About 57.95 per cent of the employed were female labourers of which 7.39 per cent and 50.55 per cent was constituted by family and hired female labour, respectively. About 0.96 per cent labour came from children. Out of this, 0.09 per cent came from family and 0.87% from hired labour.

On the whole charaka reeling units employed 2,442,12 man days/ year. In that, 63.87 per cent were family labourers and remaining 36.13 per cent was hired labour. These units employed about 56.81 per cent of male labour; of this, 46.98 per cent came from family and 9.83 per cent from hired male labourers. Totally 42.16 per cent of female labourers worked in these units of which about 16.67 per cent were family members and 25.48 per cent were hired. About 1.01 per cent child labour was involved of which 0.21 per cent came from family and 0.80 per cent was hired labour (Table 1).

Activity wise involvement of labour force: About 2442.12 man-days were employed in all the *charaka* units per annum. In this, family labour contributed more (1,559.8 man-days) than hired labour (882.32 man-days). Coming to activity wise labour force involvement, purchasing of cocoons and sale of raw silk was attended by only family males (334.73 man-days). Cooking, reeling and changing of water in the pans were attended by 827.28 man-days of family males (578.00 man-days) and females (249.28 man-days). In addition 636.55 man-days of hired labourers (139.41 man-days of men, 495.86 man-days of women) were also involved. Wheel operation was attended by 326.88 man-days of family labour and 143.96 hired labour. Collection of dead pupae, jute and feeding of fuel was attended mainly by hired labour (101.81 man-days) and by family labour (70.91man-days) (Table 2)

Table 3 indicates the percent wise involvement of men and women in different operations of silk reeling. About 87.4 per cent females and

6.4 per cent of males were involved in collection of dead pupae, jute extraction and feeding of fuel; about 50.9 per cent of females and 49.0 per cent of males were engaged in cooking, reeling and changing of water in the pans. Only 28.4 per cent of females and 68.9 per cent of males were involved in wheel operation but none of the females were involved in purchase of cocoon and sale of raw silk, where as about 100 % of males were engaged in these activities.

DISCUSSION

As recognized by the policy makers, creation of employment is the prime component in the implementation of several rural and semi-urban developmental programmes. Some of the features of sound employment generation are providing for eligible adults, reasonable security of tenure, fixation of working hours, provision of working environment, providing minimum wages, contribution towards provident fund and so on. Silk reeling industry performs a two-fold function, i.e. on one hand it can bring to the light the hidden potentialities of the rural industrial sector, which are in the shape of traditional skills and crafts. On the other hand, it can offer remunerative employment opportunities to absorb the growing surplus labour in the rural and semi-urban areas. But still the employment generation potential of reeling industry and the adoption of principle of employment have to be empirically ascertained. In this context, an attempt was made to estimate the quantum of employment generated, to comprehend the activity wise involvement of labour force in the different sized *charka* silk reeling as well as to know the fulfilments of above employment features.

Silk reeling industry being a cottage industry of excellence needs highly labour intensive activity. On an average it creates 1,022.47, 2,711.43, 3,882.38 and 2442.12 man-days of employment per annum in small, medium, large and overall *charaka* silk reeling units, respectively. It is interesting to note that more employment has been generated for family labour in small *charaka* units (78.01 %) compared to medium *charaka* units (76.44 %), large *charaka* units (28.32 %) and overall *charaka* units (63.87 %). This revealed that, in traditional technology *(charaka)* more family labour is involved mostly owing to limited resources. The proportion of family men and women labour engaged was more in the small *charaka* units compared to other categories of *charaka* silk reeling units. The proportion of hired labour engaged was very high in the large *charaka* units (71.61 %) compared to overall *charaka* units (36.13 %), medium *charaka* units (23.56 %) and small *charaka* units (21.99 %), thus clearly indicating scope for employment. These findings are in line with Kerutagi (1999). In the hired labour segment, female workers were engaged at lower wages even though they were prompt in work. Considerable number of reelers engaged number of children for various reeling operation, which is against the principle/guidelines of employment. This highlights the need for creating awareness about child labour.

Apart from the purchase of cocoons and sale of the raw silk, cooking, reeling and turning of wheel operations were prime operations in the *charaka* silk reeling; other activities were collection of dead pupae, extraction of jute and feeding of fuel. Reeling involves cooking of cocoons, collecting filaments after brushing and finally passing through a thread guide device on a large wooden wheel. The other important activity in the *charaka* silk reeling is turning of the wheel sequencially where more number of family members were employed. More hired women were also assigned for this activity.

Reeling operation, besides being a skilled activity is also a risky job as it involves the handling of cocoons at a high temperature. Labourers are exposed to noxious fumes emerging from the cooking cum reeling pans and also to sulphur dioxide released from the firewood/saw dust. Wheel operation, though did not require any skill, is straining for some children. In addition they were exposed to bad smell and smoke while collecting dead pupae. To provide appropriate health care, health clinics and welfare measures are already programmed by Developments in these directions. Effective implementation of these schemes needs attention and monitoring.

SUMMARY

The study was conducted in different *taluks* of Karnataka in respect of employment potential in *charaka* silk reeling units. The *charaka* reeling units provide jobs for male, female and children for either family or hired labourers. Silk reeling industry is a cottage industry of excellence and highly labour intensive. On an average it creates 1,022.47, 2,711.43, 3,882.38 and 2442.12 man-days of employment per annum in small, medium, large and overall *charaka* silk reeling units respectively. On the whole *charaka* reeling units create employments for about 1388 man days (57 %) in favour of male population, 1030 man days (42 %) in favour of female population and another 25 man days (1.00 %) for children per annum. The important activities involved are: purchase of cocoons and sale of raw silk, cooking, reeling, changing of water in pan, wheel operation, dead pupae collection, jute extraction and feeding of fuel etc. Wheel operation is largely carried out by male while more than 50 per cent of the women are involved in cooking, reeling and changing of water in the pan (50.9%), dead pupae collection, jute extraction and feeding of fuel (87.4%). The purchase of cocoons and sale of silk is totally carried out by male members of family (100.00%). About 57 per cent of the male population are involved in various activities, irrespective of the size of the *charaka* reeling units, women involvement was 42%, whereas only 1 per cent of children were involved in reeling activities. *Charaka* reeling units have ample potential to provide jobs to the rural folk and need strategies and developmental plans/programmes for revitalizing in order to promote traditional silk reeling in the state.

Table 1. Pattern of employment in different sized charaka reeling units n= 120

Items	Small charaka reeling units	Medium charaka reeling units	Large charaka reeling units	Overall charaka reeling units
Family labour				
Male	586.75 (56.38)	1524.48 (56.22)	808.90 (20.83)	1147.34 (46.98)
Female	208.31 (19.96)	541.22 (19.96)	287.19 (7.39)	407.33 (16.67)
Children	2.62 (0.25)	6.82 (0.25)	3.61 (0.09)	5.13 (0.21)
Sub total	797.68 (78.01)	2072.52 (76.44)	1099.70 (28.32)	1559.8 (63.87)
Hired labour				
Male	58.65 (5.73)	166.70 (6.14)	796.02 (20.24)	240.21 (9.83)
Female	158.56 (15.50)	450.66 (16.62)	1962.83 (50.55)	622.36 (25.48)
Children	7.58 (0.74)	21.55 (0.79)	33.83 (0.87)	19.75 (0.80)
Sub total	224.79 (21.99)	638.91 (23.56)	2782.68 (71.61)	882.32 (36.13)
Pooled				
Male	645.4 (63.12)	1691.18 (62.37)	1594.92 (41.08)	1387.55 (56.81)
Female	366.87 (35.88)	991.88 (36.58)	2250.02 (57.95)	1029.69 (42.16)
Children	10.20 (0.99)	28.37 (1.04)	37.44 (0.96)	24.88 (1.01)
Sub total	1022.47 (100)	2711.43 (100)	3882.38 (100)	2442.12 (100)

Note: Figures in parenthesis indicate percentage

Table 2. Activity wise employment pattern in charaka reeling units n=120

Type of operation	Family labour (man - days)				Hired labour (man - days)				Total	% to total		
	Male	Female	Children	Sub Total	Male	Female	Children	Sub Total		Male	Female	Children
Purchase of cocoon & sale of raw silk	334.73	0	0	334.73	0	0	0	0	334.7	100.0	0.0	0.0
Cooking, reeling & changing of water in the pans	578	249.28	0	827.28	139.41	495.86	1.28	636.55	717.4	49.0	50.9	0.1
Wheel Operation	223.48	99.39	4.01	326.88	100.8	34.15	9.01	143.96	324.3	68.9	28.4	2.8
Dead pupae collection, jute extraction & feeding of fuel	11.13	58.66	1.12	70.91	0	92.356	9.46	101.816	11.1	6.4	87.4	6.1
Total	1 147.34	407.33	5.13	1559.8	240.21	622.366	19.75	882.326	1387.6	56.8	42.2	1.0

Table 3: Activity wise involvement of male, female and chidren in silk reeling (Per cent wise)

Activities	Male	Female	Children	Sub Total
Purchase of cocoon and sale of raw silk	100.0	0.0	0.0	100.0
Cooking, reeling and changing of water in the pans	49.0	50.9	0.1	100.0
Wheel operation	68.9	28.4	2.8	100.0
Dead pupae collection, jute extraction and feeding of fuel	6.4	87.4	6.1	94.9
Total	56.8	42.2	1.0	1.00

Referneces

Anon. 1990. *CSR&TI- Improved charaka*. CSR&TI, Bangalore. P.14.

Anon. 2002. Survey of reeling enterprise in Karnataka. *Ann.Tec. Rep.* State Department of Sericulture, Government of Karnataka, pp. 12-20.

Kerutagi, M.G. 1991, *Production and marketing of silk cocoons in Bijapur district, Karnataka. An economic analysis*. M.Sc. (Agri.) Thesis, University of Agricultural Sciences, Dharwad. 140pp.

Sonwalker, T. N., B. N. Lakshmipathaiah & Y. V. Nagabhusanaiah. 1987. Different systems of reeling and their influence on the quality of silk, *National Workshop on silk Industry in India: Ancient Traditional to 21st Century*. CSR&TI, Bangalore. *pp. 1-16.*

40

Role of Women in Conservation of Natural Resources in Forests A Case Study from Chikmagalur District, Karnataka

LALITHA P. N AND A. K. CHAKRAVARTHY

ABSTRACT

Women of sixteen villages in Bhadra wildlife sanctuary were interviewed for assessing the role of women on forest policies and conservation of natural resources. Survey conducted during 1996-1997 indicated that women in general declined to comment upon natural resources. Ninety percent of women had no knowledge about natural resources. Except for Fire Protection Act women had no knowledge of other government policies. They had no say in policies and conservation of natural resources in Bhadra wildlife sanctuary. As a result there has been encroachment of forest area, cultivation in forest land, shifting cultivation, hunting, smuggling, open cast mining and deforestation. To make role of women active and effective for conservation of natural resources in Bhadra wildlife sanctuary, suggestions have been advanced in this paper based on a survey.

INTRODUCTION

Natural resource refers to assets supplied by nature (Somani, 1992). As per modern concept resource does not refer to a thing or a substance alone, but to a function that a thing or a substance may perform. As the world population is increasing, there is increasing consumption of natural resources like forest, fish and wildlife. Much of the Chikmagalur district in Karnataka is rich in natural resources particularly forests. There has been encroachment on customary rights and resettlement of tribes and other people living in wild forests. A study was conducted in Bhadra Wildlife Sanctuary, Chikmagalur to asses the role of women in conserving natural resources.

MATERIAL & METHODS

Chikmagalur (12° 54′ 42″ and 13° 53′ 53″ North latitude and between 75° 04′ 46″ and 76° 21′ 50″ East Longitude) though a separate district, phytogeographically is an inseparable part of the hilly region of Karnataka and the Western Ghats. This hilly, *malnad* tract is full of rich forests and is renowned for its coffee plantations. Chikmagalur is situated roughly in the south western part of the state (Karnataka). A large area of this district is largely forested hilly region of heavy rainfall.

The Muthodi or Bhadra wildlife sanctuary or Muthodi forest covers 492 km^2 and is divided into (1) Muthodi wing and (2) Lakkavali wing. The Survey was confined to Muthodi wing only. The forests in Chikmagalur district are divided into five divisions viz., (1) Chikmagalur (2) Muthodi (3) Mudigere (4) Aldur and (5) Kadur.

The forest area of Chikmagalur division is 1094.62 Km^2 out of which, 220.95Km2 area was handed over to private persons and Bhadra wildlife Divisions, Chikmagalur. There are six classes of forests in the district. They are:-

(1) State forest (29, 070.63 ha) (2) Reserve forest (86.26 ha) (3) Tiger Reserve (8666.60) (4) Sandal forest (46.68) (5) Village Forest (6) Protected forest (24392.15 ha) and (7) C and D class land (23956.37 ha) (The Mysore Forest Mannual, 1957).

The Sanctuary consists of some of the dense and most useful forests in the country. There is continous stretch of valuable forests encompassing the whole of the Jagara valley and most of Koppa and Mudigere taluks. It encompasses the hillside densely and the trees give shelter for coffee plantations.

A survey of women in the villages of Muthodi Wildlife sanctuary (Bhadra Wild life sanctuary) was conducted during 1996 and 1997 with a proforoma containing 60 questions in the local language, Kannada. In all, 60 villages were surveyed in the sanctuary from June to November 1996-97. Each field visit was of three days duration. It was conducted using a four Wheeler. Women were randomly choosen for interaction so that they represented all sections of people of village, taking into account economic status, culture, age, sex and profession. Women were contacted in groups (2 to 5) and individually. Forest department officials were not involved during interaction as they would have influenced the responses of women during such random, informal talks.

Women were contacted without prior notice in front of their houses, on roadsides, shops, schools, official building, forest plantation, bazaars, community meetings, etc. The hypothesis of the study was that women contribute to conservation of natural resources in forests so that natural resources of forest are better managed.

RESULTS

It was found that out of 125 women contacted only 18 responded. Most of the women folk in the wildlife sanctuary were reluctant to answer the questionnaire as they were illiterate, unaware of the policies, the values and conservation of forest resources. In general, women folk in the Muthodi forest area were mostly confined to indoors, shy, illiterate and not much exposed to the matters/aspects related to forests, wildlife, their rights and governmental policies.

Of the 18 women interviewed 7 were *lingayat* (upper cast), 4 belonged to schedule castes/tribes, 6 were *gowdas* (Cultivators; both backward and forward castes included) and one was a Brahmin. None of them were educated upto 7th standard. They belonged to the economic group that earned less than Rs 5000/- per month. Women were poor, devoting most of their time to meet the basic needs of the family like food, shelter and clothing though their lives revolved in and around nature in the region. They were unaware and largely ignorant of their status, pattern of exploitation and their rights over natural resources. Seventy eight percent of the women were of the opinion that forests in wildlife sanctuary were fully conserved. Twelve per cent of women did not respond to the question. Ten per cent were of the opinion that natural resources in forests are not conserved. They attributed the reasons to encroachment, cutting of trees, destruction of habitat and forest fire during summer. Women held the view that forest was over exploited from the past 25 years and no efforts were made to improve the rejuvenation capacity of the forest. One woman aged 18 of Muthodi village pointed out that timber extraction from wildlife sanctuary is the main reason for destruction of original forest. From November to May every year two lorry loads of timber per day via Sirsave and 5 lorry loads ply per day via Joldal-Chikmangular route and timber is taken away. There is no permit to cut and it is illegal to cut trees. But sandal and rosewood are cut down and the wood is illegally taken out. Miss Malathi of Malandur village was of the view that the forest department releases marginal lands, rocky, stony and dry patches for growing trees.

In response to the extent of forest conservation, 60% of women believed that 75% of forests are conserved, 25% of women held the opinion that 50 % and 75% of forests are conserved. Rest of the women (15%) were unaware of the conservation status.

Interestingly all the responded women were aware of the period and measures to contain fire in the forest. From the response it can be concluded that protection of forest from fire is the most popular programme of the forest. In Bhadra/ Muthodi wildlife sanctuary, fire in the forest persists between February and May every year. A network of informers has been established to monitor the spread of fire. Bisons and elephants run away from forest and raid crops, whenever forest fire occurs. Insects, millipedes, earthworms, birds nests, peacock, young ones of the birds and small animals like deer, wild goats, porcupine, hares, squirrels, lizards, and snakes get burnt. Teak, rosewood and bamboos are the trees earliest to get affected by fire. According to one estimate losses due to forest fire cannot be estimated but about 15 crore rupees loss occurred due to outbreak of fire in 1994 alone in the sanctuary.

Women participated in training programmes conducted by the forest department every year. On forest fire, all women answered in affirmative when questioned about fire path, effects of fire and efforts of forest department to mitigate affected area. Planters in and around the sanctuary including women co-operate and participate in extinguishing fire in the forest. Girls in schools inside the forest participate and plant trees during first week of August as a part of *Vanamathotsava* celebration. Girls (upto XII standard) and women also participate in motivation camp and Nature camp. Eighty percent women confirmed that forests are useful to them as it provides wood, land, water and good environment. It meets basic necessities including fire wood. However, seventy percent of women sampled held the opinion that the forestry resources are not easily available. Thirty percent of women asserted that forestry resource are being smuggled out of the sanctuary.

By 2005, most of the residents of Bhadra wildlife sanctuary have been rehabilitated outside the sanctuary around Bhadravathi. This policy of the government will accelerate the pace of conserving natural resources in this sanctuary. The ordinance of the Government of India to rehabilitate people living in forests (not the few tribals) outside the forests will prove useful in conservation of forest and natural resources.

CONCLUSION

Women of 16 villages in Bhadra Wild life sanctuary do not have adequate knowledge on forestry resources. This is because 84% of women have difficulties in meeting their daily requirements. A majority (> 90% of women) had no time to even think about the status of their forestry resources and their Rights, the women has no say and control over the policies of forest management which are ecologically very important for sustenance of forest and wildlife. Implementation of Fire Protection Act is effective. Questions on Wildlife Conservation Act indicated biotic interference. Thus it can be concluded that women in Bhadra wildlife sanctuary have passive role in conservation of forestry resources. Awareness camps, education, active participation and empowerment of women will help conserve natural resources.

ACKNOWLEDGEMENTS

We owe a deep sense of gratitude to officials of the Forest Department, Chikmagalur and Bangalore, to residents of Bhadra wildlife sanctuary, Center for Development and change, University Grants Commission, New Delhi, and Directorate of Collegiate Education, Bangalore for encouragement and help.

References

Mysore Forest Manual. 1957. Director of Printing, Stationary and publications, Government Press, Bangalore..456pp.

Somani, L. L. 1992. *Dictionary of Ecology and Environment*. New Delhi, 1170pp.

41

Role of Women in Agricultural and Allied Activities

SHILPARANI, M. S., Y. KATTEPPA AND V. SHIVAYYA

ABSTRACT

The total Indian population is 1027.01 millions of which women account for 495.73 millions (2001 census). Women constitute about 48.27 percent of the population. The labour force participation of the women is 51.6 percent. In rural India, agricultural and allied activities employ as much as 89.5 percent of the female labour. Women have an active role and extensive involvement in livestock production, forest resource use and fishery processing. They contribute considerably to household income through farm and non-farm activities as well as through landless agricultural labours. They are deprived of exposure to knowledge and hardly participate in the process of decision-making. Female agricultural labourers are the poorest section of Indian society. Despite the women's role in agriculture there has been a lack of adequate attention towards gender issue in the area of agriculture research.

42

Women in Agriculture New Perspectives

BALAKRISHNA GOWDA

ABSTRACT

Role of women in agriculture is a well-known fact and is unquestionable. The changing trend in agriculture technology, use of modern tools, and wider participation of community is crucial in making farming activity as an attractive field of employment. While most of the young and male population is shifting to urban areas for wage based non-agriculture activity, women have to bear the burden of agriculture in villages. The drudgery faced by women in farm activiies need to be reduced and value addition to the farm products is required at home scale/community scale in the villages. Newer and economically viable systems need to be introduced at village level with ample scope for women to participate in these activities. Women should not only be partners in agriculture but they also should take lead in entrepreneurship and in value addition. Further marketing systems for the products need to be established.

Annexure - 8(B)*

Consultation on "Hindu Succession (Amendment) Act, 2005 and its impact on Rights to Land Ownership for Women in the context of Increasing feminization of Agriculture" held on 21st September, 2005 at MSSRF (M.S. Swaminathan Research Foundation, Chennai)*

1. It is heartening to note that the Parliament has passed the Hindu Succession Amendment Bill, 2005 and the Act has come into effect from 6th September 2005. This in turn is of great relevance in the context of NCF's (National Commission on Farmers) terms of reference which interalia cover "Recommendation to be made for the credit, knowledge, skill, technological and marketing empowerment of women, taking into consideration the increasing feminization of agriculture and the proposed conferment of right to land ownership".

2. In the First Report accelerated programme activity were recommended to be taken up for (i) Existing policy pronouncements and directives of direct transfer of land to women (ii) Improving inheritance rights of women to land by

* *Extract from Third Report of National Commission on Farmers-": Saving farmers + Saving family: 2006 year's Agricultural Renewal, Dec.2005 (www.kisanyog.gov.in)*

bringing changes in laws and working towards better implementation (iii) Enabling women to buy land from the market. In the interest of social justice as well as well-being of the families, enhancement of women's rights and entitlement to lands deserves to be the first priority on the agricultural agenda.

3. The Hindu Succession Amendment Bill, 2005 addresses access to private land through inheritance. One of the most significant Amendments in the 2005 Act is deleting the gender discriminatory section 4(2) of the Hindu Succession Act (HSA), 1956. Section 4(2) exempted from the purview of the HSA significant interests in agricultural land, the inheritance of which was subject to the devolution rules specified in state level tenurial laws. But in several states the tenurial laws specify inheritance rules that are highly gender unequal. The 2005 Act brings all agricultural land on par with other property and makes Hindu Women's inheritance rights in land legally equal to men's across states overriding inconsistent state laws. Secondly it provides daughters including married daughter's birth right in Joint family property. If implemented in letter and sprit the act will have widespread ramifications for women's access to agricultural land.

4. The consultation was convened to discuss the implementation of the HSA amendment Act, 2005 in particular and women's access to land in general. It was attended by renowned subject experts, civil servants, civil society representatives and NGOs. It was expected that the consultation would focus on:

a) The adequacy of the provision of the Hindu Succession (Amendment) Act 2005 for land rights for women.

b) The steps required for effective implementation of the provision of the Act by the States.

c) Difficulties foreseen in the implementation of its provisions both in letter and sprit.

d) States requiring special attention.

e) Issues relating to supportive services like credit, technology, knowledge and skills for farm women.

f) Any other issue relevant for converting conferment of land rights into an effective tool for empowerment of women in agriculture in rural areas particularly for on-farm/off-farm income generation.

5. The salient observations and recommendations that emerged on diverse issues related to women's land rights are summarized below:

i) There cannot be a unique model for solution of the problems prevalent in different areas. Different models are needed for different areas.

ii) The voluntary transfer of land rights of women should be there. Often a woman wants to give land to her daughter, but she is restrained from doing so because of mafia/criminal pressures who simply grab her land.

iii) In most villages women get married within 7-8 km distance and they can take care of the land of their parents. Institutional mechanism to implement the law by passing the obstructions should be thought out. Without protective mechanisms, it will be difficult to implement the Act.

iv) It is essential that similar rights for cultivation be provided to married and unmarried women. Right to cultivation should be determined by being a resident of the village. If a married woman comes back for some reason she should get access to land.

v) The Hindu Succession Act should be renamed India Succession Act and be available to any citizen of India on demand. The option should be available to women from minority communities also to appeal under the act.

vi) Bureaucratic will is needed to address gender inequality. Creating a conducive environment to implement the act honestly is also necessary. The initiative of local officials in the implementation of the act should be publicized for greater adoption by others.

vii) MPs/MLAs should set an example by applying the act to themselves. It should be also a part of the service condition of IAS/Central Government employees, panchayat members etc. Government should also provide incentives to local officers to proactively implement the act. This will promote acceptance by the citizen at large.

viii) With the Hindu Succession (Amendment) Act now in place, it is necessary that the government fixes time frame for its implementation.

ix) A change in mind set is needed to address the issues facing women. There have been instances that the training has been provided to men folk to write their will so that no land is left intestate (Without a valid written will), enabling women to demand their rights. Training for writing of will that benefit women folk should be provided.

x) Ceiling law and definition of the family must be made uniform across the country. The importance of the act when the male members of the family migrate was emphasized.

xi) In the case of surplus land transfers, steps should be taken to avoid poor women getting poor quality of land.

xii) The cost of land registration should be waived or reduced when the land is registered in the name of women or a group of women.

xiii) There should be a control on leasing out practice because if the land is leased out to women farmers or group farmers then they do hard work to develop the land, get better productivity out of it only to find the owner's right can be upheld in a court of law.

xiv) There is an order from the State Government that the wasteland should be distributed to women's groups by the panachayats but the order is not having the desired effect.

xv) There is a need to start registration for land with joint pattas and it will be good if men will go for registration. State Government should document the wasteland, which is distributed under joint pattas. A wide campaigning is needed to promote joint pattas.

xvi) The ceiling surplus land should be provided to women first, and then in joint pattas and remaining land could go to men.

xvii) At present data on ceiling surplus land, land acquired by the government and land distributed by the government land under litigation is available for ten states only. There are no records available on how much land is registered in the name of men and women in agriculture. There is need to collect and computerize the data regarding this aspect and without any bias. There is need to continue consultation with state governments and try to get data on women farmers at Tehsil levels. Data should be gathered to capture the fragmentation of land holdings.

xviii) There is a need to go for Cadastral survey in the North East especially Meghalaya.

xix) There is a need to operationalise the law with greater awareness and opportunities to take maximum benefit of it.

xx) There is an urgent need for legal literacy to women in particular and farmers in general about the act. There is a strong need to strategise how this can be done and start legal literacy from school level. Legal aid should be available to the women. Small NGO group that are working in this area, Farmer's organizations, bankers and panchayats should play a proactive role in the awareness and implementation of the Act. KVCs can also play an important role in this.

xxi) Groups of women should be formed to access land accompanied by support services to improve the economies of scale because it is difficult to manage the land by individual women as she cannot access the services. Therefore, often she sells the land. There is a need to develop village fund for women's groups to purchase land for agriculture and other developmental activities.

xxii) 80% of the SHGs are women SHGs but virtually none of them are land based. Necessary action should be taken to promote land based women SHGs.

xxiii) There is a need for crop diversification and a consolidated approach for off farm and on-farm activity to enhance the income and productivity of the farm as well as assurance of livelihood and nutrition security at household level.

xxiv) Lack of access to finance is a major issue in rural areas. Inability of land banks to provide credit in spite of legislation to the effect in place should be addressed. Loans should be provided at concessional rates where land is in the name of women. Need for changes in collateral system of banking have also to be examined.

xxv) A holistic approach towards allotment of land to women should be adopted. The government should also help in developing the land, other basic facilities and soft loans (e.g. Malaysia).

xxvi) There is a need for strengthening training for farm women and engendering of training institutions. Training on entrepreneurship development should be made a part of the Agricultural University curriculum.

6. Some other suggestion in this regard would be:

(i) Ensure the recording of women inheritance shares by the patwaries, in all land records and in cases where women own land (via any means), ensure that their names are entered in the corresponding land record.

(ii) Promote schemes which foster women owned group assets, including but not restricted to, village common lands and village forests.

(iii) Provide insurance coverage to farmers for crop failure caused by lack of rain, drought, and damage by wild life and land slides.

(iv) Through media raise awareness of relevant legislation on women's right to land and resources.

(v) To ensure the institutionalization and codification laws for tribal communities.

Annexure - 8(C)*

Women in Agriculture – What Next?

1. A two-day workshop was organized by the M S Swaminathan Research Foundation (MSSRF), Chennai, on November 11 and 12, 2005, on the theme: Women in Agriculture in India: What Next? The aim of the workshop was to discuss the status and problems of women in agriculture in India ten years after the fourth World Women's Conference held in Beijing in 1995, in the light of the commitments made by the Government of India at the Beijing conference and later in the Common Minimum Programme of the UPA Government. About 25 participants drawn from government, academia and civil society, including Chairman, National Commission of Farmers (NCF) and Member Secretary, NCF participated.

2. The workshop benefited greatly from a review paper presented by Ms. C.P. Sujaya, IAS (Retd.) and Visiting Fellow in Gender and Development for 2005 at the MSSRF, who had earlier been commissioned to undertake such a review. In her paper entitled, Beijing + 10: Women in Agriculture in India, Ms. Sujaya dwelt on the three themes of growing feminisation of work, of agriculture and of poverty in the last decade, the problems in women's access to land and other assets, and lacunae in

policies and schemes, supported by data and case studies. She also summed up, in conclusion, the policy changes required to address these effectively. A succinct presentation by Prof. M.S. Swaminathan, Chairman, National Commission on Farmers and Chairman, MSSRF, provided guidance to the workshop participants on areas where concrete recommendations could be made.

DISCUSSIONS

3. The consensus that emerged, noted that issues concerning Indian Women ten years after the Beijing Conference of 1995 must be seen in the background of the consequences of the implementation of neoliberal macroeconomic policies since 1991, as well as the overall path of economic development since Independence. The structure of the economy inherited at the time of Independence and the nature of polices followed since then by successive governments at the Centre and in most States had, by and large, failed to address adequately the deeply entrenched inequalities of class, cast and gender. Often, they had contributed to the widening of inequalities. The status of women in India even before the neoliberal reforms began therefore left a lot to be desired.

4. The reforms, which consist, in the main, of policies of liberalization, privatization and globalization (LPG for short), were highly deflationary, and especially disastrous for agriculture and the rural economy, as also for key social areas such as health and education. Rural women, engaged mostly in agriculture, primarily as labourers and to a lesser extent as cultivators, were especially badly hit. While the reforms have been lauded by some as having ushered in an era of 6% annual growth rate of GDP, these figures provided poor consolation to rural workers, as the rate of employment fell sharply between 1993-94 and 1999-2000 as compared to the period 1987-88 to 1993-94. Agriculture and the rural economy bore the brunt of neoliberal policies of removal of all quantitative restrictions on imports, steep lowering of import duties on agricultural produce, slashing of input subsidies, collapse of institutional credit, near-absence of public investment, reduction in rural development expenditures, weakening of the public distribution system and

decline in allocations for agricultural research and extension. The rate of growth of agricultural output in the post-reform period stands at less than 2% per annum, while the rate of growth of food grains output has fallen below the rate of growth of population for the first time since Independence during this period. These trends have been well-documented elsewhere and taken note of in the earlier Reports of the National Commission on Farmers.

5. It is against this disastrous performance of agriculture and the rural economy that the problems facing women in agriculture have to be understood and discussed. Women have been negatively impacted upon by these polices which have added to the problems women already face in a deeply patriarchal social milieu with structured gender inequality along economic, social, political and cultural dimensions. The collapse of rural and agricultural employment under neoliberal policies has led to large-scale migration, often (through not always) sex-linked. The migration of male earners has led to a certain degree of feminization of agriculture, as well as to a significant proportion of female-headed household. This in turn has meant that a disproportionate share of the agricultural crisis is borne by women. In many States of India, property and inheritance laws have traditionally discriminated against women, and despite some steps taken in recent years to redress the situation, women are still mostly to be found recorded as agricultural labourers rather than as cultivators. The traditional patterns of community land and forest management in many tribal areas (Schedule 5) and especially in the hill States of north-eastern India (Schedule 6 and other areas) have entirely different property and inheritance laws, and here too the brunt is borne by the women. This is of course partly also due to the non-recognition of the work women do on their family farms.

6. The decline of institutional credit for agriculture and the onerous terms on which farmers are having to borrow are major factors contributing to rural distress. At the same time, during the last decade, there has been a highly visible micro-credit initiative, directed mostly through SHGs of women. The growth and experience of SHGs across the country is highly variable and region specific. While in some parts of the country, there has been a phenomenal disbursement of funds to women in rural

areas, and some capacity building, via the intermediation of NGOs, State agencies, line departments etc., which has enabled them to meet some pressing consumption needs, it is far less clear to what extent this has enabled access to improved technology, higher productivity and profitable enterprise. SHGs in themselves can and certainly do provide women a new and legitimate social forum where they can come together and explore avenues of collective action, but such instances have been only part of the story.

7. On the one hand, there is evidence that already overburdened rural women have now to shoulder even more responsibility and work, while the withdrawal of the State in crucial areas of rural infrastructure and welfare is justified in the name of empowering women and reducing the dependency culture. On the other hand, in some places, where the agriculture sector has been depressed and offering limited opportunities for work, the SHG movement has provided some diversification, off-farm employment and greater awareness of technological inputs. The conclusion that emerged recognized the value of SHGs as a space for women's agency, but also noted that a great deal of capacity building of women in SHGs was still required to enable them to truly empower themselves. It was also pointed out that the promotion of SHGs and the provision of micro credit can only be a supplement to direct measures which provide women land and other productive assets, individually or in groups, that can serve as collateral for larger-scale, productive loans.

8. The conditions of existence of women in rural areas have been further worsened by the withdrawal of the state from education and health. Women now have to manage even more stringent family budget constraint in the face of increasing costs of health care and education. Evidene of greater violence against women in various forms points to the impact of the rural economic crisis in a context of embedded patriarchy as well as the rampant consumerism promoted by the forces of globalization.

9. The emergence of an active land market, dominated by speculators, especially in rural areas adjoining urban centers has led to encroachment on, or outright destruction of, common

property resources, where again poor and dalit rural women are among the worst sufferers. They are now not only having to trek longer distances to fetch water, fuel and fodder and to attend to their ablutions, but also find their productive activities adversely affected.

10. While neoliberal policies had caused great distress to the rural economy in general and rural women in particular, this period has had some positive developments as well. Important among these were the emergence of Panchyati Raj Institutions (PRIs) and peoples' movements of various kinds, including the SHG movement and wider movements questioning the logic of LPG policies and seeking a more people-centred policy frame. The workshop participants felt that both PRIs and SHGs needed considerable resource support for capacity building and that the processes of capacity building should be strongly engendered to strengthen and empower rural women. A stronger and more effective interface is needed between PRIs at all levels, particularly their women members and the *Shag* and their federations, wherever they exist.

11. The issues of land reforms and land rights for women were both important, and the State should play a crucial role in the gendered implementation of land reforms, wherever relevant, since in almost the entire north- east hill States it is not the State but village communities that control land. Preventing the erosion of the few traditional systems wherever these continue to exist where authority and agricultural decision-making is shared by women and men, is also a goal to be kept in mind.

12. The question of land lease is tricky, but lease arrangements that empower poor rural women without allowing land grabbing by more powerful forces including corporate should be the desirable approach. In this context, it is noteworthy that the Common Minimum Programme (CMP) of the UPA Government provides:

i) "The UPA government should ensure that at least one-third of all funds flowing into panchayats should be earmarked for programmes for the development of women and children, village women and their associations should be encouraged to assume responsibility for all development schemes relating to drinking water, sanitation, primary education, health and nutrition".

ii) "Complete legal equality for women in all spheres should be made a practical reality, especially by removing discriminatory legislation and by enacting new legislation that gives women, for instance, equal rights of ownership of assets like houses and land".

13. Apart from removing discriminatory legislation and new legislation, legal and civil society supports are required to ensure that new rights are not inherently biased against community group rights and against non-land holding individuals-women, dalits, and the poor.

14. The recommendations that were made at the meeting are presented under the following heads: Land and Productive Assets, Labour and Employment Generation, Common Property Resources and NRM, Support Services, Self Help Groups, Data Gathering, Women in Technology and Research.

RECOMMENDATIONS

15. *Land and other productive assets:* Ideally, both men and women should be recognized as heads of the household, and all property be jointly held, though provision would have to be made for equal sharing of all property in the event of cessation of marriage. While this may be a long-term goal, in the present context, it is necessary to urgently take steps to reduce the burdens and disadvantages faced by women due to lack of title to property, especially land, which excludes them from access to resources and inputs, particularly land. This is particularly true with reference to the banking system and its structural limitations, which require proof of title as collateral. Hence it is recommended that:

i) From a gender perspective, because of women's multiple productive roles, it is necessary that agricultural policy and programmes adopt a farming systems approach that integrates agriculture, livestock, fish, forestry and water resources, instead of treating these as separate sectors as at present.

ii) All new assets which have accrued to the household (before a specified cut-off date) by any means (purchase, transfer, grant etc) should be registered in the name of both husband and wife, applicable to all assets such as land, houses, trees, animals, equipment etc.

iii) The above should also apply to membership of groups/ categories which are prerequisites for access to resources, e.g. water users' associations.

iv) Banks should be asked to accept spousal ownership/ membership as collateral for loans and extend Kisan Credit Card to women.

v) The distribution of land mandated by Government of India in the 1980s with regard to surplus land, wasteland, and ceiling surplus should be monitored and recorded and up to date records prepared within a specified time limit.

vi) Lands, particularly wastelands, vested with government should be transferred to women's groups (including SHGs) for productive use and appropriate economic activity.

vii) The implementation of land reforms in a gender-sensitive framework should be closely monitored and up to date records prepared, keeping in mind recent legislation (Amendment to Hindu succession Act 1956). The issue of community-held land has to be separately addressed.

viii) Given the failure of successive attempts to ban swidden (*jhum, podu*) cultivation in which women are especially involved, it is necessary to develop and diversify swidden with multiple and high value crops (medicinal, aromatic plants) to increase diversity, enhance food security and perhaps improve women's income.

ix) Resource support, value addition and market linkages for the traditional crafts in which farming women are involved, either full time or as supplementary activities, need special attention.

16. *Labour and employment generation:* Wage employment is still and continues to remain the most important source of income for the rural poor, especially women, and hence the primary need is to revive and strengthen capital investment in agriculture and rural infrastructure, which will generate farm and non-

farm employment. In this context, the NREG Act as well as six major schemes-National Horticulture Mission, Bharat Nirman (New Deal for Rural India), National Rural Health Mission, Krishi Vigyan Kendras in all districts (ideally to be re-designed as Krishi our Udyog Vigyan Kendras), setting up of SHG Capacity Building and Mentoring Centres (as recommended by the NCF), and establishing women-managed Community Food, Water, Fodder and Feed Banks–provide immense scope for employment of poor rural women. Equally important, the new policy arrived at after a series of Round Tables of state ministers and secretaries for Panchayat Raj have arrived at new modalities for disbursement of funds to panchayats, and for the planning process, which is now to be initiated at the level of Gram panchayat and built up from below. Major central schemes are now to be channeled directly to panchayats. Safeguards have been built in to ensure that the funds are not delayed in payment.

i) There should be adequate representation of women in all bodies at Gram Panchayat, Block and District levels in all bodies concerned with generating, planning, designing and implementing employment. For example, (a) Standing Committee on Planning of the Panchayats at village, Block and District level (b) Technical support groups set up at each level (c) All Standing Committees at the three levels. In the sixth Schedule, areas where the 73rd Amendment does not apply, the engendering of traditional institutions at the village level should be encouraged and supported

ii) There should be similar adequate representation of women in all bodies concerned with training and capacity building at village, Block and District levels–(a) in setting up of training and capacity building institutions and (b) in planning the content and methodology of training.

iii) Forty per cent of all employment generated through new schemes should be reserved for women.

iv) Similar reservation should be made for the support services arising out of new employment opportunities for women to ensure not only that women's services are utilized, but that their labour is recognized and their knowledge and skills built up.

v) Increased emphasis should be placed on implementation of Minimum Wages, Equal Remuneration and other existing laws.

17. *Common Property Resources and NRM:* Increasing illegal encroachment on common property resources (village grazing lands, ponds etc) with the approval of the local authorities, PRIs and traditional community authorities, who are often governed by partriarchal values and may not be gender sensitive, has severe consequences for the poorest and most marginalized, especially women, who depend on CPRs for livelihood. So does disposal of wastelands to commercial interest. In order to prioritise livelihoods of the poor/women over commercialization it is recommended that:

i) To discourage local authorities, Gram Panchayats or equivalent body from commercial disposal of village CPR and wastelands, they should be required to place such issues before the Gram Sabha for decision, in order to ensure that voices of marginalized sections (women, dalits and poor) are not excluded.

ii) Awareness generation should be enhanced at all levels, and especially among the bureaucracy and local levels, traditional and elected bodies, to support poor women/marginalized groups in their struggle to protect their existing use rights over CPR.

iii) Institutional and funding support for the formation of women producer's associations to process, transport and market farm produce, milk, fish, crops etc. should be provided.

iv) Commercial interest/companies should be banned from acquisition of wastelands for purpose of direct cultivation. CBOs (women's group, Dalit groups, tribal groups, SHGs) should be given priorty for acquisition, lease, or grant of wastelands for cultivation.

v) Since submergence of ecologically important watersheds, pastures and agricultural lands through hydro-electric schemes in the hills has a far-reaching impact on women's livelihoods in both the upland and downstream areas, run-of-the-river schemes should be preferred to reduce displacement of people and erosion of agro-biodiversity.

vi) Monopolies in the NTFP trade in different States and for different produce should be reduced with a view to their ultimate elimination and to ensure fair prices to the producers.

vii) Different kinds of social networks of women producers (or of women and men) that already exist should be strengthened to enhance their ability to negotiate with traders and others.

viii) The question of individual vis-a-vis common property resources needs to be carefully studied and addressed so that public policies do not erode systems of common access and use even as they ensure individual rights, both of which are important.

18. *Support services:* The various social services provided by the State & community groups, such as health services, child care services, early childhood education, midday meals, ration shops are generally perceived, both by the concerned departments and the public at large as welfare measures. Their role as support services for poor women is less understood, and hence close linkages are not established, nor the programmes adequately structured to be flexible in addressing these needs in a context-specific manner. Hence, it is recommended that:

i) ICDS should be redesigned to become more context and gender sensitive to the needs of poor working women, especially in the rural areas, by (a) its decentralization and management by PRIs and (b) enabling local managements to modify the timings, location, provision for care of younger children, child worker ratio to suit local needs (c) providing additional fund sources such as NREG/ powers to raise additional funds to meet such context-specific needs.

ii) Innovative proposals to provide crèches for young children (below three) and cooked meals on site for women labour should be worked out in the case of NREG work-sites, which can be adapted elsewhere.

iii) Identity cards for migrant women workers should be provided to enable them to access PDS, secure admission to primary school/anganwadi/etc for their children.

iv) The social security provisions in the two Acts before Parliament relating to unorganized workers should specifically include all categories of women workers in the agricultural and rural sector.

v) Common facility (meeting place/workshed) for women should be provided from the Gram Panchayat Mahila Kosh (as already proposed in the First Report of NCF, Chapter IV on New Deal for Women in Agriculture).

19. *Self-Help Groups:* The tremendous expansion and visibility of SHGs in the last decade and a half has brought many gains, both social and economic, to poor rural women, though there are marked regional differences. There is debate about the extent and nature of gains, as increasing burden of work and responsibility on women, increased domestic violence, limited economic advancement, continuing social and caste oppression, and constraints to decision making in the home have also been observed. While recognizing the tremendous employment potential of SHGs, the continuing dependence of poor women on multiple sources of livelihoods implies that there are no panaceas, and multiple solutions have to be sought. At the same time, empowerment of women at the political, community, family and social levels is emerging as a precondition for further gain. It is hence recommended that:

i) Critical and region-specific studies of the SHGs drawing on available studies and data should be taken up to look into the financial, economic-productive, social-community, political, family-household, and personal levels of women's empowerment.

ii) Based on such studies, a more detailed understanding of the place of SHGs in women's multiple livelihoods may be built up, as well as mapping the location of women in the rural and agricultural sector.

iii) The banking system should be urged and enabled to develop a broad structure of varied financial services to support the multiple livelihoods of rural poor women, using the SHGs as the basic organizational structure.

iv) A network of capacity building institutions should be set up to strengthen and develop SHGs to undertake the various functions/activities into which they are expanding, including ToT, and to nurture them during process.

v) A series of training modules should be developed and delivered through various mechanisms including AIR and mass media, distance education providers, village Knowledge Centres with the help of ICT.

vi) Awareness creation through mass media, people's organizations, and networks should be used to create visibility, not only for SHGs and their achievements, but for the role of women in the economy.

20. *Data gathering and use:* The importance of gender-disaggregated data in all fields and areas cannot be over-emphasised, not only in order to create visibility for the role of women in the economy, society and policy, but also as a tool for planning, monitoring and evaluation. It is equally important that in future all records and documentation of all types are gendered, for the same purposes. The endeavour should be to make the process of data-gathering participatory and to include qualitative measures of development and poor women's lives and livelihoods. Measures of gender equality including the gender disaggregated data collected and collated through the different recommendations, should not be an academic exercise but be incorporated into agricultural and related development policies. Hence, it is recommended that:

i) Time use studies should be extended to all the States, and should be repeated at regular intervals. These will strengthen the process of creating visibility for women's unpaid labour and domestic work.

ii) Discussions with the national data-gathering agencies should be initiated by the concerned Ministries (Ministry of Agriculture) to continue and strengthen the process of gender-disaggregated collection of basic information through the decennial Census, the National Sample Surveys, and other occasional and regular data collection surveys of all Departments.

iii) Gender-disaggregated data on vital statistics should be collected through District level surveys, which would enable estimates of district level IMR, U5MR, MMR, JSR, and other important development indicators.

iv) The next Agricultural Census should collect gender-disaggregated data on operational holdings.

v) To begin the process of estimating women's contribution to GDP, pilot studies at district level should be initiated to work out the methodology for such calculation.

vi) The responsibility for collecting gender-disaggregated information on vital statistics should gradually be shifted to PRIs, after ensuring that (a) the elected leaders as well as the concerned functionaries are sensitized to its importance and the methods needed and (b) that the necessary funds and facilities are transferred to them.

vii) Institutions must be sensitized to the need to maintain records of the representation of women in all committees.

21. Skill Empowerment and Technology for Women

21.1. Technology for women requires special attention and the impact of major on-going schemes (and any future schemes) on the knowledge and skills of poor. Rural women engaged in farming and allied activities need to be systematically studied. The lack of tools designed with women in mind is yet another area of concern. With the objective of empowering women in all fields of environmental management, including water harvesting, wasteland development, sustainable agriculture and livestock development, biodiversity conservation and its sustainable and equitable use, ongoing schemes should be "engendered" on priority basis. The National Commission on Farmers was requested to develop a strategy and set up mechanisms to advise on "engendering" four major programmes where women's concerns should need attention. These are:

- National Horticulture Mission
- Bharat Nirman Programme
- Capacity Building and Mentoring Centres for SHGs
- Revitalisation of Krishi Vigyan Kendras

21.2 All of these programmes need to be assessed in order to see firstly, how the present design impacts on poor women and secondly, whether and how these could be used to strengthen women's capacities. Skill development of women will be a major focus, and the engendering of these programmes can offer a guideline for the inclusion of women in other technology-based or technology-rich programme areas.

22. *Research- The NCF:* The NCF chairman's suggestion to set up a national level action and policy research network to carry out longitudinal studies of women's roles in agriculture and rural livelihoods in the various agro-ecological regions of the country was strongly endorsed. The network can be on the lines of a Hub and Spokes model, with the hub centre performing servicing, facilitating and integrating roles, and mobilizing the power of partnership for women's empowerment in agricultural and rural development.

Summary and Recommendations of National Symposium on Women in Agriculture, Bangalore

INTRODUCTION

Agriculture in India accounts for 37% of country's GDP employing 70% of working population and about 84% of all economically active women (GOI, 1991). Not only these women play their traditional role of home maker, care giver, mother, wife and but they also work along with men to keep the farm viable. However, the gender bias in agriculture system forms a big obstacle to achieve the goal of sustainable agricultural productivity because of the heavy burden placed on women. Compared to men, women have limited access to land, credit, extension services, secure employment, rural organizations and political bodies. They do not get paid for their role as family managers, are paid comparatively less for on farm work, have less time for self care, have no role in decision making or spending the money earned by them, have no voice in policy decisions and planning of developmental activities, receive almost no gender sensitive training in agricultural production in spite of their greater involvement and worst of all they are the largest, unorganized workforce in the country. The National

Symposium on Women in Agriculture held at the University of Agricultural Sciences, Bangalore during April 10-12, 2007 was a forum to debate these issues. The participants were physicians, economists, sociologists, veterinarians, biotechnologists, agronomists, plant breeders, entomologists, extension specialists, nutritionists, farm women, agricultural officers, NGOs involved in rural development etc.

CONSTRAINTS IDENTIFIED IN THE LIFE OF WOMEN IN AGRICULTURE

(1) *Traditional role:* Women are traditionally home makers and care givers. Apart from household management, child care and social responsibilities, rural women are unpaid workers on the farm either as labourers or helping their men folk in rising, harvesting and storing crops leaving little time for effective participation in local organizations. They have limited opportunities to articulate their needs and concerns during planning process in the local bodies because of men's dominance.

(2) *Gender bias in accessing resources:* Traditional perceptions and attitudes perpetuate gender biases and discriminate against rural women's access to community and household resources. At the same time degradation of natural resources in the process of economic growth is threatening women's access to subsistence livelihood resources in rural areas. Globalization could further marginalize rural women and result in greater inequity in access to resources. Lack of or poor education may adversely affect her as new technologies emerge in agriculture and information sector.

(3) *Mechanization and modernization of agriculture* have led to increased agricultural productivity and decreased drudgery but have adversely affected the women from lower castes, lower income groups, and landless families or with small holdings by reducing demand for labour. Further, mechanization has occurred for activities usually carried out by men and women continue to toil in labour intensive chores. Introduction of weedicides and pesticides has deprived a large number of women labourers of their meagre source of income.

(4) *Lack of or inadequate education* is a key contributing factor to gender bias in accessing resources. Women's education is vital for their

empowerment, to develop basic skills needed to participate in knowledge intense agriculture and economic activities.

(5) *Technology development and dissemination* have not been responsive to household drudgery associated with routine activities of rural women at home and on farm. In most technologies, women's actual needs are ignored. In fact, implements designed to increase production and reduce manual labour have increased the workload of women. Appropriate technologies should be developed to consider not only women's workload, but also the potential effect of the technology on their status and the economic control over resources and property.

(6) *Credit*: The provision of collateral free, group guaranteed micro credit has provided an important source of capital for rural women. These group based micro credit approaches have served the short term financial needs of rural households well, improving cash flows, benefiting women's livelihood, leading to social solidarity and changing social status. However there are negative effects by way of increased social conflicts in the community and the tendency of men to control women's access to economic assets, women being used as the front for control over loans or financial decisions. Expansion of micro credit is limited by loan size, poor access to opportunities to improve skills as they are based on only existing skills, cost effective technology and knowledge of market complexity. The repayment schedules (compulsory weekly repayments plus contributions to savings) and interest rates are not compatible with long term agricultural production and livestock rearing cycles. This has led to women in rural areas to pursue additional income generating activities to keep up with repayment schedule by increasing their workload.

(7) *Global and regional factors affecting rural women:* Globalization, economic integration, commercialization, urbanization, advances in agricultural and information technology, political instability, HIV/AIDS, livestock epidemics and natural disasters have serious effects on rural women. Their impact is perceived on agricultural productivity, economic vitality of villages, household food security, health and economic security of families. Nonetheless advances in agriculture and information technology as well as economic integration provide opportunities too for the livelihood of rural women and work. However, HIV/AIDS, political

instability and natural disasters exert additional pressure on rural women's access to resources and their work as farmers and producers.

(8) *Economic transformation and agricultural contribution:* Usually regional economic growth is accompanied by a steady reduction in the relative share of agriculture to Gross Domestic Product (GDP). The decreased agriculture affects rural women adversely. The negative impact comprises loss of livelihood, unsustainable food production, increased dependence on cash economy to access food, search for alternate, viable livelihood, etc. In this scenario, limited skills, low or no education render rural women more vulnerable. With agriculture being replaced by export driven production, land resources get consolidated forcing subsistence farmer to sell his land. In the process both rural men and women become agricultural labourers from being farmers.

(9) *Global economic integration:* The demand for agricultural commodities in the global market acts as a force that can influence production and affect rural women's life depending on the type and scale of farming. Demand for cash may drive women to sell their home-grown produce, small livestock and home made food in local markets and urban centres. However, such home-originating enterprises are likely to face competition from large scale, industrially processed food products.

(10) *Environmental destruction and natural disasters:* Economic growth is mostly accompanied by destruction of environment with negative consequences for farm women. Environmental degradation can be chronic or sudden. Chronic degradation is mostly due to deforestation which in turn is a result of forest clearance for agriculture, logging and fuel wood collection. Sudden degradation of environment is due to natural disasters and industrialization. Both affect rural women's life adversely.

(11) *Limitations of agricultural extension system:* The agricultural extension system conceived and operating to train and capacity building of farmers is almost exclusively composed of men and the attempts to reach out to women farmers is highly inadequate. Further, most of the transfer of technology is conceived with male farmer as the target. The time and location of training hardly takes into consideration the time and work pattern of farm women

(12) *Biotechnology:* There is general lack of information about the extent to which women farmers in the region know about and understand the potential opportunities and risks of agricultural revolution brought about by biotechnology. Given the poor educational background of rural women and the requirement of highly sophisticated involvement in biotechnology based agriculture, it is likely that the majority of rural women would be further marginalized by such development. On the other hand, there is a need for in depth studies on the taste, palatability, acceptability and long term effects of biotech foods.

(13) *Information and Communication Technology (ICT)* has great potential to inform and educate rural women about agriculture, health and nutrition provided they are user friendly, have gender responsive content and applications. However, there are barriers to increased use of ICTs such as inadequate infrastructure, high cost, limited capacity and content relevancy.

(14) *Occupational health hazards of farm women:* Very few studies have examined work-related injuries among farm women. Compared to this there is at least some data on pesticide effects. Still there is a need for understanding the gender implications of pesticide use. Because in addition to the general implication of pesticides in leukaemia, lymphoma, aplastc anaemia, soft tissue sarcoma and cancers of various kinds, a variety of reproductive health problems of women are linked to pesticide exposure. These include miscarriages, stillbirths, delayed pregnancy, increased risk of birth defects, disruption of oestrus cycle and disturbance in menstrual cycle. Very low levels of pesticides have been found to induce DNA damage, mental retardation and delayed milestones in children born to mothers exposed to pesticides. Pesticide risk is higher to people suffering from asthma, diabetes and cardiovascular diseases.

RECOMMENDATIONS OF THE SYMPOSIUM

(1) *Detailed data* on the time allocation to different domestic and farm activities, access, ownership and control of assets and resources, their potential, constraints and limitations are essential for economic analysis and policy decision in agricultural planning, rural development and reform in agriculture. There is

a need to collect sex disaggregated data and gender differentiated information for all aspects of agriculture and rural development. Such information enables the understanding of contributions, resource deprivations and helplessness of rural women and in formulation of programmes and policies for their betterment. Such a database also helps in assessment of progress made in the status of women in agriculture. This needs to be done across the country where socio-economic conditions are vastly different from region to region so as to evolve area specific agrarian planning and reforms.

(2) *Drudgery Reduction:* Improvement and popularization of tools reducing drudgery such as sitting type groundnut decorticator, dibblers for sowing, dehusking and pearling of ragi etc. are to be popularized and made available at affordable price.

Rural electrification facilitates access to water, fuel and use of kitchen appliances reducing drudgery.

Alternate energy sources to firewood such as gobar gas, solar power etc are not only sustainable but economic and reduce the drudgery of fire wood collection.

Basic services need to be expanded in the villages.

(3) *Empowerment of women through Self Help Groups:* Apart from helping to set up enterprises based on the needs, aptitude of farm women, availability of resources and marketing potentiality of the area, the members of SHGs should be trained in skills and project management.

(4) *Post harvest technology and creating marketing outlets:* Adequate facilities to train women in post harvest technology specially fruits, vegetables and flowers need to be addressed. Lack of transportation facilities, adequate storage space, price fluctuations and lack of planning in sales are other problems perceived.

Improved infrastructure to facilitate marketing and links with urban commercial centres are needed.

(5) *Strategies for empowerment:*

- *Ownership and access* to land, forest resources, pastures, irrigation and water should be facilitated.
- *Availability of easy credit*

- *Capacity building by development of technology kits and mixes* for promoting knowledge and skill empowerment on various aspects of farming, household technology and to improve the quality of farm produce.
- *Diversification of economic activities* in rural areas in an interdependent and integrated manner at micro level within the limits of available resources.
- *Need for an integrated management service* at the village level for effective utilization of various services and inputs received by rural women.
- *Outreach programmes* through radio and TV.
- *Improving decision making capacity*
- *Easy availability of farm equipment, supply of farm inputs* at subsidized rates by the government
- *Design of low cost implements, and provision of crop insurance.*
- *Rural electrification* facilitates knowledge acquisition through radio and TV, provides entertainment, communication, reduces drudgery of women through use of appliances, and acts as a resource for production.
- *Training in professional management i.e.* management of assets, enterprise and organization building
- *Linking up with environmental and consumer groups*
- *Access to government agencies* and developmental organizations
- *Collectives of women* can pool resources including labour, share experience, knowledge and skills and develop capacities to deal with power structures and systems
- *Child care services* viz establishment of farm crèche and provision for supplementary feeding to infants and toddlers in the crèche
- *Girl's education* to be promoted by offering low cost education and establishment of schools in all villages. There is a need for effective, need based education for rural women.
- *Adult literacy programmes,* skill development and training

- *Health campaigns* with regular check ups, preventive and theraptic approach, self care and proper rest and sanitation facilities
- *Agro-processing and value addition to the farm produce* not only increase the nutritive quality of food but also increase income of farm women and creates job opportunities
- *Development of nutritional guidelines*, establishment of nutrition gardens to provide cost effective micronutrients.
- *Propagation of peri urban horticulture based on tissue culture* propagation or "micro propagation" as a cottage industry.
- *Production of bio control agents* as an income generating venture
- *Adoption of villages* using Participatory Rural Appraisal techniques for sustainable, self reliant and people oriented development activities

(6) *Redesigning and Extension interventions* to suit the requirements of women. These programmes *should be practical, short-term, should use audio-visual inputs* and be scheduled at the right time of the day and close to the residences of target women. The interventions will be more effective if information relevant to women and local problems is generated by research and the training is imparted by women extension officers. The impact assessment of training and follow up of trainer's performance are important.

It is necessary to *link agricultural extension system to groups of women farmers* so that women's needs and priorities become part of mainstream policies and agricultural planning.

A collective review of and revision of content and the methodology of teaching by extension workers and farmers both men and women will be a useful tool in capacity building of both the teacher and the trained. It may be beneficial to have biannual orientation of extension functionaries towards gender sensitization.

(7) *Biotechnology:* There is large demand for tissue cultured ornamental plants, banana and sugarcane. Scope for tissue culture based peri urban horticulture for women is great for several reasons. The National Horticulture Board readily funds establishment of tissue culture units. Added to this, women with

their intrinsic qualities for delicate work are ideally suited for this endeavour. With little training and financial assistance from NHB, an averagely educated woman can get trained, start a unit and gainfully employ other women.

(8) *Ensuring sustainable use of natural resources* is possible only if policy makers, planners and development agents have a clear understanding of the relative and shifting roles of men and women in natural resource management i.e. division of labour, access to resources, decision making, traditional knowledge and practices.

(9) *Pesticide Hazard:* Data is needed on the ways in which women are exposed to pesticides in agricultural production and also the differential pattern of pesticide use between men and women, unique health impacts of pesticides on women, awareness of women about the extent of pesticide impact on health etc. There should be increased awareness about sustainable agriculture amongst women, emphasis should be on reduced use of agricultural chemicals. Women's groups, farmers and extension functionaries should be trained in alternative, non chemical ways of controlling pests and diseases. Integrated pest management should be propagated widely. Organic farming also reduces pesticide risk. Legislation on pesticides should be strictly enforced. There should be training of women on pesticide usage or such training should be a mandatory component of all agricultural training programmes. Development of pesticide action kit is a must.

(10) *Promoting gender-sensitive planning and policies :* The planners and macro economists in various development agencies should identify gender biases impeding development in the status of rural women . Advocates of gender equity and policy makers should work in tandem to achieve economic development.

There should be a gender responsive and women inclusive curriculum for students of agriculture and rural development courses.

Technologies useful to farm women should be evolved based on gender sensitive participatory research methods. Participatory components have to incorporated in to on going research programmes such as plant breeding, natural resource management, storage of food grains, integrated pest management, water shed programmes etc.

(11) *Decision making:* Studies on the extent of deciosion making by farm women are warranted as very few studies have been undertaken about women's involvement in decision making in rural families in spite of tremendous involvement of women in agriculture, livestock management and post harvest activities.

(12) *Valuation of unpaid work:* An appropriate technique to value the unpaid work of women in the home and on farm is required and attempts be made to reward such work. Women's unpaid work should be considered while planning development programmes.

(13) The emphasis of the national meetings on rural women should move from activities connected with emancipation to making strong propositions on macroeconomic policies globally, regionally and nationally. Women's voice is deplorably missing on issues like foetal killing of female child, high rate of mortality of girls, protests against establishment of Special Economic Zones in agricultural lands by displacing marginal farmers, fate of family members surviving suicides by farmers, plight of pavement vendors, tiny enterprises by rural women, the mobilization of members of SHGs to produce small consumer goods instead of sustainability in agricultural production, in framing agricultural policies in the context of emphasis on growth of GDP even at the cost of agriculture, hunger and food security amongst the rural poor, in labour legislation where it is intended to replace unionized labour by flexi labour and political-economy debates. Thus there is an urgent need for a co-ordinatation amongst the various forums devoted to betterment of rural women to identify the impact of globalization, liberalization, rapidly changing rural life, decline in agricultural productivity etc and to design both short term and long term remedial measures.

Reference

Government of India, 1991. National Sample Survey. 38th Round. Natl. Sample Survey Organization, Ministry of Statistics and programme implementation, GOL, New Delhi.

Subject Index

A

B

C

MIX
Papier | Fördert gute Waldnutzung
FSC® C083411

Zeitfracht Medien GmbH
Ferdinand-Jühlke-Straße 7
99095 Erfurt, Deutschland
produktsicherheit@kolibri360.de